Message from the President

77 years of WAZA – this is an appropriate date to have a look back onto our history in order to realise where we have come from and how successfully our organisation has developed during these years!

Starting as a prestigious gathering of zoo directors WAZA is now the worldwide voice of zoos and aquariums, representing the highest standards in animal keeping, conservation action and education. The goals of zoos have changed over time and WAZA is reflecting that.

Many decades were needed to collect sufficient experience on how to successfully keep and breed animal species from all over the world. Today only zoos and aquariums have the knowledge and capacity to keep sustainable populations of hundreds of threatened species to safeguard their survival for the future.

WAZA and its members have a unique history and an even more demanding future. Let us not underestimate the role that we can play for species, habitats and humans!

Jörg Junhold, WAZA President 2012–2013

Contents

Preface 6
Introduction 8
A Timeline of WAZA History 10

Chapter 1: The World Association of Zoos and Aquariums in an Evolving World 17
Humble Beginnings 18

Chapter 2: Changing Times, Changing Organisational Structure 25
The Evolution of Membership 26
The Evolution of the Constitution and Bylaws 33
Organisational Nuts and Bolts 38
The Changing Face of WAZA 45

Chapter 3: The World Association of Zoos and Aquariums' Commitment to Conservation ... 61
WAZA's Evolution into a Conservation Organisation 62
WAZA's Relationship with IUCN 101
The International Studbook Programme 112

Chapter 4: The Evolving Role of the Zoo 119
The Evolution of Zoos 120
Environmental Education 142
The Heini Hediger Award 150
WAZA's Role Today and in the Future 155

Appendices 161
Appendix I: WAZA's first members (1935/1936) 162
Appendix II: WAZA's Code of Ethics and Animal Welfare 163
Appendix III: WAZA's 2010 Bylaws 168
Appendix IV: Locations of WAZA's Annual Conferences from 1935 to 2012 179
Appendix V: 'What Does Our Union Stand For', 1952 paper by Armand Sunier 180
Appendix VI: 'The Changing Role of Zoos in the 21st Century', 1999 paper by William Conway ... 184
Appendix VII: 'Conservation of Nature: A Duty for Zoological Gardens', 1964 paper by Kai Curry-Lindahl 192

Index 197

Preface Gerald Dick, WAZA Executive Director

When WAZA started to consider and work on this commemorative volume, the question was put forward about the importance of such an undertaking, as well as the year to choose for the celebration. It was soon felt that putting together historical facts after more than 70 years of existence is worth pursuing and that a 'round number' anniversary is not that exciting – that is why 77 years was chosen. This decision also reflects a modern and more marketing-driven understanding of the global community of zoos and aquariums. Looking back and getting historical facts right will lead to a better understanding of the roots of the Association and related developments over time. Furthermore, it will help give a better understanding of the present situation and future developments. This publication covers the very early days of international cooperation between zoos and aquariums up to and including the year 2011. I would sincerely like to thank the WAZA Council for the support of this project, especially Gordon McGregor Reid, the WAZA President at the time when the decision was taken. I also wish to thank Laura Penn, the first author of this publication, who undertook a search through the WAZA archives, analysed old documents, and put everything together in an accurate and easy-to-read form. Thanks are also due to Markus Gusset for editing and critically reviewing the text, compiling the illustrations and document reprints.

Gerald Dick

Co-authors Laura Penn (top) and Markus Gusset

The main sources of information for this publication were the minutes and proceedings of the Annual Conferences as well as the brochure '25 Years International Union of Directors of Zoological Gardens 1946–1970' published by Walter Van den bergh in 1973. Thanks to IUCN librarian Katherine Rewinkel El-Darwish, additional conservation-related literature could be located.

In order to make such a project happen, the support of numerous individuals and organisations is necessary and has to be acknowledged.

Acknowledging the financial support of the following WAZA founding members or current members: Basel Zoo, Bristol Zoo, Bronx Zoo/Wildlife Conservation Society, Budapest Zoo, Chester Zoo, Cologne Zoo, Copenhagen Zoo, Frankfurt Zoo, Houston Zoo, Kaliningrad Zoo, Leipzig Zoo, Munich Zoo, Nuremberg Zoo, Paris Zoo, Skansen Foundation, Smithsonian National Zoological Park, Vienna Zoo and Zoological Society of London.

Acknowledging the comments on a draft of the text by: Tim Brown, William Conway, Lee Ehmke, Gunther Nogge, Mark Penning, Alex Rübel, Christian Schmidt, Sally Walker and Roger Wheater. Fiona Fisken of the Zoological Society of London kindly helped with edits and comments. Gerhard Heindl, the professional historian of the Vienna Zoo, not only undertook the very first perusal of the WAZA archives but also remained an invaluable advisor throughout this project of documenting the history of WAZA. Photographs were kindly made available by the individuals and organisations indicated as the copyright holders in the legend accompanying the illustrations. Additionally, I would like to especially thank: Jörg Adler (Münster Zoo), Geneviève Beraud-Bridenne (Muséum national d'Histoire naturelle Paris), Catherine de Courcy (Dublin Zoo), Felix Kirschenbauer (Munich Zoo), Julie Larsen Maher and Marion Merlino (Wildlife Conservation Society), Peggy Martin (Lincoln Park Zoo), Klaus Michalek (Naturschutzbund Austria), Manfred Niekisch (Frankfurt Zoo), Lex Noordermeer (Rotterdam Zoo), Christian Schmidt (previously of the Frankfurt Zoo), Christoph Schwitzer (Bristol Zoo) and Elizabeth Townsend (IUCN/SSC Conservation Breeding Specialist Group).

Acknowledging the contributions to the text by: Rick Barongi, Leobert de Boer, Onnie Byers, William Conway, Peter Dollinger, Lee Ehmke, Nate Flesness, Jenny Gray, Stephan Hering-Hagenbeck, Jörg Junhold, Willie Labuschagne, Robert Lacy, Joanne Lalumière, Lena Lindén, Julia Marton-Lefèvre, Colleen McCann, Stephen McKeown, Dave Morgan, Manfred Niekisch, Gunther Nogge, Peter Olney, Mark Penning, George Rabb, Alex Rübel, Karen Sausman, Mark Stanley Price, Simon Stuart, Sally Walker, Chris West, Jonathan Wilcken and Shigeyuki Yamamoto.

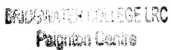

Introduction

Julia Marton-Lefèvre, IUCN Director General

With more than 300 member organisations across the planet, WAZA has become the leading voice of the worldwide community of zoos and aquariums which are 'United for Conservation'. The International Union for Conservation of Nature (IUCN) and WAZA have been on a long conservation journey together. IUCN is proud to have WAZA – at the time called the International Union of Directors of Zoological Gardens (IUDZG) – among its founding members and today as our major partner and neighbour at the IUCN Conservation Centre in Gland, Switzerland.

Very early on, IUCN recognised the part that modern zoos could play in achieving conservation goals. Many WAZA members have played an active role in the IUCN Species Survival Commission and, in particular, in its Conservation Breeding Specialist Group. This collaboration culminated in 1993 with the publication of *The World Zoo Conservation Strategy*, which defined, for the first time in a single document, the role of the international zoo and aquarium community in supporting nature conservation.

Indeed, WAZA members have been instrumental in key conservation successes. Several species which had been extinct in the wild, including the Arabian oryx, California condor and Przewalski's horse, have now been reintroduced by zoos and wildlife parks to their original habitat. More recently, the WAZA community made a great effort to respond to the amphibian extinction crisis by supporting many field conservation projects.

Over the years, WAZA has spearheaded many significant improvements in the living conditions of animals, and contributed towards increasing our knowledge of the biology of threatened species. Thanks to WAZA, cooperation between zoos has greatly improved, making them a much more powerful tool to achieve conservation goals.

Finally, with more than 700 million visitors passing through zoos and aquariums each year, WAZA members are the single biggest gateway for raising public awareness about conservation – a key objective of the United Nations Decade on Biodiversity (2011–2020). For many urban dwellers, it is often the only opportunity to see up close the amazing diversity of life on our planet, and learn about its condition and the reasons why it should be preserved. Many zoos and aquariums are already promoting the use of the continually updated *IUCN Red List of Threatened Species* in their public-outreach efforts.

As is the case with all organisations that have a long and rich history, IUCN and WAZA have had to reform and reinvent themselves along the way to become the truly modern global conservation networks that they are today. In doing so, both have built on a simple but powerful idea: by working together, we can achieve more.

On behalf of the entire IUCN family, I warmly congratulate WAZA on its 77th anniversary. Today, more than ever, IUCN and WAZA stand United for Conservation.

A Timeline of WAZA History

1935
- First meeting of the 'International Association of Directors of Zoological Gardens', Basel, Switzerland
- The Association's Constitution was approved on 1 October
- Kurt Priemel elected as first President of the Association

1946
- Official re-establishment of the organisation after World War II, renamed the 'International Union of Directors of Zoological Gardens' (IUDZG), Rotterdam, the Netherlands
- Armand Sunier elected as President
- IUDZG seat in Paris, France
- US$ 4 membership fee

1947
- IUDZG Constitution approved
- Members resolved to save species from extinction

1948
- International Union for the Protection of Nature (IUPN, now IUCN) founded, with IUDZG as a founding member, Fontainebleau, France

1949
- IUDZG joined IUCN as a member

1952
- Concern for live-animal trafficking voiced by members
- 'What Does Our Union Stand For?' paper presented by Armand Sunier, Rome, Italy

1954
- Members unanimously accepted a decree calling for the restriction on the export and import trade of wild-caught birds
- US$ 5 membership fee initiated

1956
- First logo
- Members resolved to help end the illegal trade in wild animals
- Members realised the potential of zoos to participate in captive breeding

1957
- Proposal adopted to establish an Admissions Committee for new members

1958 • IUDZG wrote a letter to the President of the 'Popular Chinese Republic' asking for measures to be taken to protect the giant panda

1960 • First support (in the form of a resolution) to an international wildlife project: IUCN's Africa Special Project

1962 • Constitution revised, San Diego, CA, USA

1963 • Resolution refusing to purchase wild-caught orangutans

1965 • Resolution on the procurement and trade of monkey-eating eagles
• Donation of £ 1,000 for an *in situ* conservation programme

1966 • Official establishment of international studbooks at the Zoological Society of London, UK

1967 • Resolution on the procurement and trade of Zanzibar red colobus monkeys, Galapagos tortoises and Aldabra tortoises

1968 • First supplementary payment of US$ 500 to IUCN's Species Survival Commission (SSC)

1969 • Two IUCN resolutions affecting zoos

1972 • First time an Annual Conference is devoted to one specific subject – education in zoos – Amsterdam, the Netherlands

1974 • Establishment of the International Species Information System (ISIS)

A Timeline of WAZA History

1975 • IUCN resolution recognising the importance of zoological and botanical gardens in the conservation of wild species

1977 • Letter written by IUDZG to the Chancellor of the Federal Republic of Austria to protect Neusiedler See

1979 • The largest expenditure to date by IUDZG – CHF 10,000 – spent on a symposium on the 'Use and Practice of Wild Animal Studbooks', Copenhagen, Denmark

1980
- *The World Conservation Strategy* published by IUCN
- The first IUCN policy statement directly related to zoos: 'Principles and Recommendations for Keeping Wild Animals in Captivity'
- The first International Studbook Coordinator designated – Peter Olney

1981
- Constitution revised, Washington, DC, USA
- IUDZG paid US$ 1,000 to assist IUCN/SSC's Conservation Breeding Specialist Group (CBSG)
- As world conservation issues were increasing and becoming more complex, IUDZG members resolved to become more active

1983 • A very conservation-minded IUDZG was emerging, Melbourne, Australia

1986 • Started granting automatic membership to successors of retired zoo directors

1987 • Battle call for the reorganisation of IUDZG – the goals and operational mode had to change in order to keep up with a changing world, Bristol, UK

1988 • First official IUDZG policy on education

1991 • Organisational milestone – IUDZG decided to accept institutions and national and regional associations as members, Singapore

1992
- Name change to IUDZG – World Zoo Organisation (WZO)

1993
- *The World Zoo Conservation Strategy* published by WZO and IUCN/SSC's CBSG

1994
- ISIS Secretariat agreed to house WZO Secretariat
- International Zoo Educators' Association (IZE) formally introduced to WZO members, São Paulo, Brazil

1995
- *Zoo Future 2005* published
- WZO sought cooperation with IZE
- WZO Secretariat opens in Apple Valley, MN, USA

1996
- First recipients of the Heini Hediger Award – Ulysses Seal and George Rabb

1997
- IZE selected to be WZO's official 'education arm'
- First Committee for Inter-Regional Conservation Cooperation (CIRCC) training grants awarded
- First Zoo Marketing Conference, Aalborg, Denmark

1998
- New name – World Zoo Organisation (without the IUDZG prefix)

1999
- WZO legally incorporated in Switzerland
- *Code of Ethics* adopted
- William Conway delivered a momentous keynote speech – 'The Changing Role of Zoos in the 21st Century' – inspiring action in the direction of *in situ* conservation, Pretoria, South Africa
- Second Zoo Marketing Conference, Amsterdam, the Netherlands

A Timeline of WAZA History

2010
- WAZA Executive Office moved into the IUCN World Headquarters in Gland, Switzerland
- Bylaws revised
- WAZA was chosen as CBD partner of the International Year of Biodiversity (IYB)
- An education manual, *Biodiversity is Life*, was produced jointly with IZE
- First WAZA book, *Building a Future for Wildlife: Zoos and Aquariums Committed to Biodiversity Conservation*, published
- WAZA received a grant of US$ 25,000 from the Mohammed bin Zayed Species Conservation Fund to support five WAZA-branded conservation projects
- WAZA received a grant of CHF 50,000 from the MAVA Foundation to support IYB
- MoU signed between WAZA and IZE, Cologne, Germany

2011
- IZE hosted by the IZE President's institution
- WAZA signed an MoU with IZE for further cooperation
- Seventh Zoo Marketing Conference, Granby, Canada
- WAZA signed an MoU with CBD in support of the United Nations Decade on Biodiversity (2011–2020), New York, NY, USA
- WAZA signed an MoU with the Alliance of Marine Mammal Parks and Aquariums (AMMPA)
- WAZA signed an MoU with the Secretariat of the Convention on International Trade in Endangered Species of Wild Fauna and Flora (CITES), Geneva, Switzerland

Chapter 1

The World Association of Zoos and Aquariums in an Evolving World

If all future meetings run as smoothly and usefully as the one in Basel [in 1935], then our Union will not only serve to promote zoo science and true colleagueship in our profession, but will also contribute modestly to a mutual understanding amongst the people.

Kurt Priemel, WAZA President 1935–1939

Humble Beginnings

Participants at zoo directors' meetings in Berlin in 1901 and 1905 (© Berlin Zoo)

Participants at animal auctions in Antwerp between 1909 and 1913 (© Kurt Priemel)

German-speaking zoo directors began meeting regularly in the late 19th century. They would assemble at the recurring animal auctions held at the Antwerp Zoo, which from 1854 took place in the spring and autumn of each year. The auctions were very popular not only with zoo directors but also with amateurs, financiers and animal dealers. The zoo directors used the opportunity of the event to exchange ideas and information with each other. Regular attendance led to the formation of a group that was the predecessor of the 'Verband Deutscher Zoodirektoren' (VDZ, Association of German Zoo Directors). At this time, the German-speaking zoo directors' meetings were informal gatherings; a legally based association had yet to be established. As Heinz-Georg Klös, Past Director of the Berlin Zoo, stated on the occasion of celebrating the 100-year anniversary of VDZ in 1987: 'Es war eher ein zwangloses Treffen Gleichgesinnter zum Austausch von Erfahrungen' (it was rather an informal meeting of like-minded people in order to exchange information). The formal establishment of VDZ took place in Berlin, Germany, in 1951. The animal auctions continued until 1913 just before the onset of World War I.

Participants at the Conference of Directors
of Zoological Gardens of Central Europe
(left column from top)
in Vienna in 1926 (© Vienna Zoo),
Rotterdam in 1927 (© Vienna Zoo),
Breslau in 1928 (© Kurt Priemel),
(rigt column from top)
Berlin in 1932 (© Berlin Zoo) and
Copenhagen in 1934 (© Copenhagen Zoo)

The Journal
Der Zoologische Garten

In order to facilitate information exchange for German-speaking zoo directors, the Frankfurt Zoological Society founded the journal *Der Zoologische Garten* in 1859. The journal is still in print today and is the world's oldest zoo-related periodical, despite a rocky history including name changes and financial constraints. In 1922, for example, the production of the journal was stopped due to a lack of funding. Strategies for rescuing and reviving the journal were popular themes at zoo-director meetings at that time. By 1928, the journal was re-established as a new series and a decision was made to accept papers in English and French in addition to the journal's original German. *Der Zoologische Garten* is the official organ of both VDZ and WAZA.

A separate and more international group of zoo directors was formed after World War I called the 'Konferenz der Direktoren mitteleuropäischer Zoologischer Gärten' (Conference of Directors of Zoological Gardens of Central Europe). The group included zoo directors from Germany, Austria, Switzerland, the Netherlands, Belgium, Poland, Denmark, Hungary, UK and Sweden. It is unclear when the first international meeting took place, but the Conference of Directors met annually from the mid-1920s until 1935 (Basel, 1925; Vienna, 1926; Rotterdam, 1927; Breslau, 1928; Elberfeld, 1929; Leipzig, 1930; Munich, 1931; Berlin, 1932; Frankfurt a. M., 1933; Copenhagen, 1934; Basel, 1935).

The proceedings from the 1930 Annual Conference in Leipzig reveal that the group discussed a wide range of issues, including whether or not they should draft 'Statuten' (a Constitution). It was felt that an official Constitution would help the group to attract more members and form a more internationally diverse group. Meeting participants also discussed whether or not membership should be individual or institution-based. Participants had lively discussions about the qualitative differences between the two types of membership. The point was made that individual membership would allow the zoo directors, who were all trained scientists, to meet and discuss issues relevant to their zoos and to the science of keeping animals. On the other hand, if institutions were invited to be members of the group, it was felt that there would be a risk that 'non-technical' staff would be sent to represent the institution, such as the financial officers and administrators, thus diluting the scientific quality of the meetings. Conference participant Kurt Priemel, Director of the Frankfurt Zoo, was especially opinionated, arguing that the status quo should be maintained wherein only zoo directors should be members of the group.

Cover of the first issue of *Der Zoologische Garten* from 1859

In 1934, Conference participant Heinz Heck, Director of the Munich Zoo, was given the task of drafting the group's first Constitution. In 1935, the Conference of Directors of Zoological Gardens of Central Europe received both a new name, the 'Internationaler Verband der Direktoren Zoologischer Gärten' (International Association of Directors of Zoological Gardens) and a Constitution, which came into force on 1 October 1935. Kurt Priemel was elected as the Association's first President for a planned term of four years. The same year, Conference participants, consisting of 18 European zoo directors, discussed the active re-establishment of several European zoos, such as the zoos in Rome and Paris, concluding that the new Association should have a broader international representation and that besides German, English and French should also be used as negotiating languages. In 1936, the Association included the following 12 countries: Belgium, Denmark, Germany, France, UK, the Netherlands, Austria, Poland, Sweden, Switzerland, Hungary and USA (Appendix I). As a result of World War II, the newly formed Association had a very brief existence. The group met only four times, in Basel (1935), Cologne (1936), Munich (1937) and Amsterdam (1938).* The 1939 meeting was planned to take place at the zoo in Rome at the invitation of Director Lamberto Crudi, but did not happen because of World War II.

Participants at the founding meeting of the International Association of Directors of Zoological Gardens in Basel in 1935 (© WAZA)

Cover of the proceedings of the 1930 Annual Conference in Leipzig

* According to the published minutes of the Amsterdam meeting, it was called the '50th Annual Meeting of the Association'. This was the fourth meeting of the newly established Association. Presumably, the German zoo directors' meetings from previous years had been counted.

Participants at the second meeting of the International Association of Directors of Zoological Gardens in Cologne in 1936, pictured at Lake Laacher (above) and in front of the Mülheim-Ruhr Aquarium (© WAZA)

Zoos at War

By Manfred Niekisch,
Director of the Frankfurt Zoo

World War I (1914–1918) affected zoos in Germany economically and in many other ways, but World War II (1939–1945) brought complete physical destruction to most zoos in Germany. Problems with heating and shortage of food caused the death of many zoo animals from the start of the war in 1939. From 1940 onwards, zoo animals considered to be dangerous in the event that their enclosures may be destroyed by war activities, such as poisonous snakes and big cats, had to be killed as a 'precautionary measure', in some cases even by direct order from the top level of the Nazi regime. When aerial bombing damaged German cities in 1943, 1944 and 1945, the zoos were also hit and most of the animals that had survived fell victim to the bombs.

Many animals not killed directly by the air raids had to be shot because they escaped from bomb-damaged enclosures or because there was no more food for them. Unfortunately, this part of zoo history is poorly documented and very little is known about zoo staff called to arms.

The maintenance of many zoos during wartime and their rebuilding immediately after the end of the war is to the credit of quite a few heroes, many of them unknown. In Frankfurt, for example, strict orders to kill all animals during the last days of war were simply not implemented by the city employee in charge when he received assurances from four keepers that they would stay on and take care of the remaining animals. Thus, when Bernhard Grzimek took over as director a few days after Germany had surrendered to the allied forces, he could build up the zoo from those remaining animals.

It is related that around the end of the war the keepers painted the bread they fed to the animals with green paint so that hungry people would not steal it. Like Frankfurt, other important zoos, such as those of Nuremberg and Berlin, started reconstruction immediately, unlike Düsseldorf where the reconstruction of the famous zoo (opened in 1876) was discussed for 35 years before the idea was finally abandoned in 1979. Instead, a modern building containing the former Löbbecke-Museum and the new Aquazoo was erected. In Frankfurt, the area around the zoo was completely devastated so Bernhard Grzimek took the opportunity in that time of confusion and disorder to simply 'add' a few hectares to the zoo area by putting up signs that he had obtained from the US Army, an initiative that resulted in several court cases and a city zoo that was larger than before the war.

Bomb-damaged Frankfurt Zoo in 1944/1945 (© Frankfurt Zoo)

World War II left many European zoos with catastrophic damage, prompting the need for increased cooperation between zoological gardens in order to rebuild. Under the leadership of Koenraad Kuiper, Director of the Rotterdam Zoo, seven European zoo directors from allied or neutral countries came together on 24 September 1946 to form the 'International Union of Directors of Zoological Gardens' (IUDZG). These included the zoos in Amsterdam, Antwerp, Basel, Copenhagen, Paris, Rotterdam and Warsaw. Zoo directors from Germany, Austria and Italy were not invited to join. IUDZG's first official language was French.

Chapter 2

Changing Times, Changing Organisational Structure

Sometimes I miss the small group of zoo directors of yesteryear. The black-and-white group photographs, the quibbling over simultaneous translations, the tie new members receive as a symbol of patriotism and our most distinguished manner of addressing one another; always saying 'Good morning Colleague', 'How are you Colleague', 'See you next year Colleague'.

*Willie Labuschagne,
WAZA President 2000–2001*

The Evolution of Membership

From the start, membership to IUDZG was an exclusive affair, reserved only for those who according to Article 3 of the 1947 IUDZG Constitution were:

'The directors, that is, the responsible leaders of Zoological Gardens or Parks managed on a scientific basis, or of important departments of such institutions, who have had a scientific education or who have proven by their work in Zoological Gardens or Parks that as members of the Union, they can be placed on a level with those who have had scientific training.'

In a speech given at the 1952 Annual Conference entitled 'What Does Our Union Stand For?', IUDZG Past President Armand Sunier, Director of the Amsterdam Zoo, stated that: 'In the eyes of the public and of the authorities, the standing, and therewith the authority of our Union, will always depend exclusively upon the quality and certainly never upon the quantity of our members'.

The first honorary member of IUDZG, Koenraad Kuiper (© Rotterdam Zoo)

Participants at the 1952 Annual Conference in Rome (© Spartaco Gippoliti)

IUDZG strictly adhered to its deep-rooted objective of representing 'scientific institutions which promoted zoological research and the protection of the world's fauna' and, as such, did not allow directors of commercial zoos nor individuals associated with animal dealerships to become members. The process of selecting members followed firm guidelines. According to the IUDZG Constitution, candidate members had to be proposed by two members. Additionally, four-fifths of the members had to vote for their admission. Great importance was given to knowing the candidates personally, with prospective candidates often being invited as guests to Annual Conferences in order for the members to get to know them better. After reading the recommendations of two sponsors and scrutinising the prospective IUDZG member's curriculum vitae, members voted for the candidate's admission.

In the early days, honorary membership was reserved for those who had rendered exceptional service to the Union. The first honorary member of the Union was Koenraad Kuiper, the person responsible for the renaissance of the Union after World War II, who retired soon afterwards as Director of the Rotterdam Zoo. A second honorary member was appointed in 1952, Lee Crandall, Past General Curator of the New York Zoological Society. In 1953, Armand Sunier was granted honorary membership. After 1953, the appointment of former members as honorary members became less severely restricted. In 1956, it was even suggested that the title should be accorded automatically to every full member who reached the end of his membership honourably. This was a controversial suggestion and, instead, it was proposed that two categories of honorary members were created: life membership, to be granted to all members retiring honourably from the Union; and honorary membership, to be reserved for all members who had made an important scientific contribution or who had rendered exceptional service to the Union.

The second honorary member of IUDZG, Lee Crandall (© WCS)

Participants at the 1950 Annual Conference in London (© Dublin Zoo)

In the end, the appointment of life members was abandoned and honorary membership was granted much more liberally than had been intended by the Constitution.

In 1957, IUDZG President Walter Van den bergh, Director of the Antwerp Zoo, proposed the establishment of an Admissions Committee, which would make a preliminary examination of the qualifications of candidate members. The Chair of the Committee would be the Secretary-in-Office and the group would consist of 'three members chosen as far as possible to take account of the geographical representation of the Union's membership so that the Committee is in the best overall position to inform itself on the merits of candidates from the various regions of the world'.

The geographical diversity of members grew modestly over the first 34 years after IUDZG's re-establishment, as did the overall membership (right).

By the 1970s, IUDZG's long-standing membership policies were starting to be questioned by some members. Operating in an increasingly international context with a limited constituency (there were only 32 members in 1970), the Union's membership policies were holding the organisation back from expanding its influence. In 1971, IUDZG members Ronald Strahan, Director of the Taronga Zoo, and Ronald Reuther, Director of the San Francisco Zoo, co-authored a document proposing 'Recommendations for a Stronger and More Responsive International Union'. They suggested that the Union was a private and exclusive club with limiting rules for membership. They further stated that the Union 'should not make claims beyond its competence [because it] did not represent the zoos of the world, nor the scientific, non-commercial zoos of the world, nor the zoos of which its members were directors'. Strahan and Reuther made a series of recommendations in their paper on ways in which the organisation could improve. They suggested that IUDZG should transform from a 'private club into a professional association'. Furthermore, they recommended that the Union should seek to establish formal relations with national and regional zoo associations, and to assist in their international cooperation in order to help promote increased international collaboration on matters affecting zoological gardens.

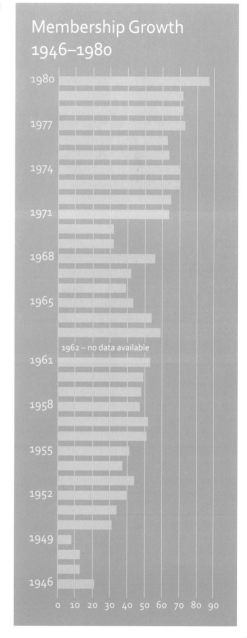

Participants at the 1971 Annual Conference in Prague (© Bristol Zoo)

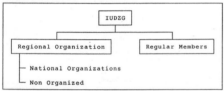

Proposed IUDZG organisational structure in the proceedings of the 1990 Annual Conference in Copenhagen

Despite the efforts of Strahan and Reuther, however, change and improvement of the Union's internal structure was slow. It was not until 1989 when their suggestions began to take shape. By then, a number of other IUDZG members had come up with similar proposals and it was felt that the organisation 'should restructure itself in such a way that for the outside world, it becomes the most representative body for the world's zoo community', as stated by IUDZG Past President Dick van Dam, Director of the Rotterdam Zoo.

By 1990, the wheels of change were in motion. IUDZG member Frederic Daman, Director of the Antwerp Zoo, chaired a working group on the 'Constitutional and Structural Changes to IUDZG'. The group recommended that membership should be changed to become dual: 'One part [should] be the regional organisations, which indirectly represent the national organisations and the non-organised zoos. The other part [should] be the regular members'.

Upon hearing Frederic Daman's report, IUDZG President Roger Wheater, Director of the Edinburgh Zoo, was from past experience sensitive to many of the members' resistance to organisational transformation, stating that: '[I know] members cherish the spirit of the Union, but times do change and the Union must change with it. There is no doubt that unless IUDZG moves with the times, then someone else will move in and take over the international functions that should be the proper right of this Union'.

Furthermore, Roger Wheater added that: 'The Union is in constant change. It is however, not the intention to bring into the Union institutions as regular members that are not satisfactory. There is not an intention to enormously enlarge the Union by hundreds of new members. It is a process of ensuring that the umbrella is in place and that the constitution underneath that umbrella allows for the kind of world role that we believe we should play. This role the Union actually cannot play as things stand at the moment'.

Frederic Daman added to the discussion by stating that: 'The move to invite associations as members was necessary as otherwise another worldwide union of national and regional associations would no doubt be founded. The Union has to change so that it is truly representative of the zoological gardens all over the world'.

Joining in the chorus of support for organisational change, IUDZG member Theodore Reed, Director of the National Zoological Park in Washington, DC, added that: 'This should have been done years ago, but on earlier occasions the delegates had not been able to make any decisions'.

WAZA Membership Today

WAZA membership has come a long way from its IUDZG roots. In addition to having a more diverse constituency, which as of December 2011 consisted of 262 institutional, 24 association, 16 affiliate, 15 corporate, and 101 life and honorary members from over 50 countries, WAZA requires that all members sign its *Code of Ethics and Animal Welfare* (Appendix II). WAZA also has more stringent membership qualifications, which require that prospective institutional members have high standards of:

- Animal husbandry and veterinary care
- Participation in coordinated species-management programmes
- Contribution to relevant scientific studies
- Compliance with national and international legislation
- Maintenance of record systems, and cooperation with studbook and species-support programmes
- Environmental education programmes
- Ethical guidelines
- Support of national and international conservation programmes
- Membership in a recognised regional and national association as appropriate

Membership Growth 2001–2011

Year	Members
2011	313
2010	303
2009	291
2008	288
2007	273
2006	261
2005	249
2004	238
2003	216
2002	212
2001	191

Cover of the membership brochure, produced in 2011

The year 1991 was a landmark year and a 'very important moment in the history of the Union', as denoted by Roger Wheater. IUDZG members voted unanimously to fundamentally change the eligibility for membership, which had existed since 1935, granting membership only to individuals who were directors of zoological gardens. The revised rules stated that:

'The following are eligible for membership of Union:

- **Zoological gardens and aquaria** who undertake to abide by the obligations of membership as laid down or as constitutionally amended by the Union and as represented by the institution's director.
- **Associations of zoological gardens both national and regional** whose members are regulated by standards approved by the Union as represented by an elected officer of the association or its chief executive officer.'

In conjunction with the breakthrough decision of changing the Constitution to include institutions and national and regional associations as members, there was a move to change the name of the Union to the 'International Union of Zoological Gardens' (IUZG). There were 48 votes in favour of the name change and 17 against. According to the Constitution four-fifths of the total votes cast had to be in favour of any amendments, so the attempt failed. The following year, however, members unanimously agreed to change the organisation's name from IUDZG to IUDZG – World Zoo Organisation (WZO) in order to reflect its new identity.

Altogether, the sweeping changes made in 1991 and the name change adopted in 1992 transformed WZO from an organisation of individuals to an organisation of zoos, aquariums and zoological associations, making it a more effective organisation with a much broader constituency and setting it on course for greater pursuits.

The Evolution of the Constitution and Bylaws

This organisation is built on beliefs, idealism and professional leadership by dedicated, mission-driven people, who share their knowledge, visions, dreams and challenges in strengthening fellowship with one another.

Peter Karsten, WZO President 1993

The International Association of Directors of Zoological Gardens, consisting of 18 European zoo directors, adopted its first 'Rules' on 1 October 1935:

1. The aim of the Union is to promote the objectives of Zoological Gardens, and also those of wild life preservation, animal fanciers and animal breeders in so far as they are allied to those Zoological Gardens.

2. Membership is personal. It is conditional on the member being actively engaged on work of interest to the Union, and lapses on his relinquishing his post or his active practice. Candidates cannot propose themselves for membership; new members can only be elected when proposed by the President of the Union. The election of new members normally takes place by vote at a meeting of the Union, but members who are unable to attend are to be asked to vote by letter. Elections, however, may also be conducted throughout by letter. A candidate shall be considered elected when at least four-fifths of the total number of members have voted in his favour.

3. The President of the Union shall be elected by secret ballot on a majority vote; he shall serve for a term of four years and shall be eligible for re-election. He shall undertake the correspondence of the Union and shall send in the Report of each meeting not later than three months after the date of the meeting.

Participants at the 2010 Annual Conference in Cologne (© WAZA)

4. Each year one member shall invite the Union to hold their meeting with him and to inspect his Gardens or premises. He shall, in conjunction with the President, prepare the programme for the meeting. He shall further have the responsibility, financial and otherwise, of reporting the meeting and shall provide the President with a shorthand report or notes as basis for the Report.

5. The language employed officially at the meetings shall be English, French or German. At any one meeting, for reasons of general convenience, that language shall be employed which is understood best by the majority of the members present.

6. In the event of serious disputes between members, each shall appoint another member as his representative to discuss the matter. If these representatives fail to reach agreement, they shall agree upon another member as arbitrator, whose decision shall be binding.

7. These rules come into force on 1 October 1935. In the first meeting held on 23 September 1935, Dr. Kurt Priemel has been unanimously elected [as President] until 1 October 1939.

Eleven years later and after World War II, seven European zoo directors from Amsterdam, Antwerp, Basel, Copenhagen, Paris, Rotterdam and Warsaw came together on 24 September 1946 in Rotterdam to form a re-established International Union of Directors of Zoological Gardens (IUDZG). Those directors who had been members of the former Association resolved that it should cease to exist because of the war. They decided to set up a new organisation, which would adopt similar rules to the previous Association, but with a few additions. 'The Union's Constitution was approved and brought into force in 1947' and subsequently 'ratified in Copenhagen on 5 September 1949':

Article 1

The title of this association, founded in Rotterdam on the 24th of September, 1946, shall be the International Union of Directors of Zoological Gardens. The Union's Headquarters shall be in Paris, France. It may be changed if so decided by two-thirds of the Union's members.

Article 2

The object of the Union is to promote cooperation between directors of Zoological Gardens and Zoological Parks managed on a scientific basis, that is, entirely non-commercial institutions with cultural and educational aims in which the public is allowed to watch and study live animals and which promote zoological research in the widest sense and also the protection of the world's fauna. The Union shall be moreover an international body representing the aforesaid institutions and promoting their concerns.

Article 3

The Union is an association of persons who are the directors, that is, the responsible leaders of Zoological Gardens or Parks managed on a scientific basis, or of important departments of such institutions, who have had a scientific education or who have proved by their work in Zoological Gardens or Parks that as members of the Union, they can be placed on a level with those who have had a scientific training.
No person shall be elected a member unless he is proposed by two members and four-fifths of the members shall vote for his admission.

Article 4

Membership ends when the member ceases to hold a responsible post in a Zoological Garden or Park or direct an important department of such an institution. Membership shall also end in respect of any person when the Zoological Garden or Park which he represents ceases, in the opinion of four-fifths of the members of the Union, to be in accordance with Article 2 of this constitution.

Article 5

On the proposal of two members the Union may, by a majority vote, grant the title of Honorary member to former members of the Union. Honorary members have the right to attend the Union's meetings and participate in discussions, but not to vote.

Article 6

The Board of the Union is to consist of at least a President and a Secretary. Members of the Board are elected by a majority vote of the Union's members. They also hold office for three years and are eligible for re-election at the expiration of those three years. In the President's absence he shall be replaced by the oldest member of the Union willing to act temporarily as President. In the Secretary's absence he shall be replaced by the youngest member of the Union willing to act temporarily as Secretary.

Article 7

Members of the Union shall pay to the Board an annual subscription of at least US$ 4. With the approval of at least four-fifths of the Union's members any member may be exempted from the annual due. Honorary members pay no annual due.

Article 8

The Union shall hold at least one meeting each year, preferably in a Zoological Garden or Park the Director or responsible leader of which is one of the Union's members. Notice of each meeting shall be mailed by the Board to all members of the Union at least one month before the meeting. Five members shall constitute a quorum at a meeting. At the first meeting of the year the Board shall submit to members for approval the statement of income and expenditure for the previous year. Minutes of each meeting shall be drawn up by the President and Secretary in the interval before the next meeting. Each meeting shall begin with the reading of the minutes of the previous meeting, making such amendments as are necessary, and subsequently confirming them.

Article 9

Except in those cases where this constitution requires a majority of four-fifths or two-thirds of the votes of all members of the Union, matters may be decided by a majority vote of all members of the Union. With the same exceptions, matters may be decided in meetings by a majority vote of members present.

Article 10

Within the terms of this constitution, supplementary regulations may be drawn up from time to time as far as this may be made necessary by meetings of the Union.

Article 11

At the request of two or more members of the Union the Board shall put forward to members an amendment to the constitution. Proposed amendments require the assent of at least four-fifths of the Union's members.

Time passed and the Union found itself in a changing administrative landscape, leading to a re-writing of the Constitution in 1962. The term 'aquariums' was added, as were a number of other adjustments dealing with membership, the make-up of the Board and the implementation of the Annual Conferences. The Constitution was changed again in 1981; thereafter, it has been officially revised five times resulting in the 2010 version, which reflects a significantly different organisation with a much broader international influence (Appendix III).

Badge of Cedric Flood, Director of the Dublin Zoo, at the 1951 Annual Conference in Amsterdam (© Dublin Zoo)

Organisational Nuts and Bolts

A chronological list of IUDZG, WZO and WAZA Presidents from 1935 to 2013
* denotes exceptional terms of office

Year	President	Institution
1935–1939	Kurt Priemel	Frankfurt Zoo
1946–1949	Armand Sunier	Amsterdam Zoo
1950–1952	Heini Hediger	Basel Zoo
1953–1955	Axel Reventlow	Copenhagen Zoo
1956–1958	Walter Van den bergh	Antwerp Zoo
1959–1961	Freeman Shelly	Philadelphia Zoo
1962–1964	George Mottershead	Chester Zoo
1965–1967	Ernst Lang	Basel Zoo
1968–1970	Wilhelm Windecker	Cologne Zoo
1971*	Charles Schroeder	San Diego Zoo
1972–1974	Zdeněk Veselovský	Prague Zoo
1975–1977	David Brand	National Zoological Gardens Pretoria
1978–1980	Colin Rawlins	Zoological Society of London
1981–1983	Heinz-Georg Klös	Berlin Zoo
1984–1986	Lester Fisher	Lincoln Park Zoo
1987–1988*	Dick van Dam	Rotterdam Zoo
1989–1991	Roger Wheater	Edinburgh Zoo
1992*	Siegfried Seifert	Leipzig Zoo
1993*	Peter Karsten	Calgary Zoo
1994–1995	Gunther Nogge	Cologne Zoo
1996–1997	Palmer Krantz	Riverbanks Zoo and Garden
1998–1999	Frederic Daman	Antwerp Zoo
2000–2001	Willie Labuschagne	National Zoological Gardens Pretoria
2002–2003	Alex Rübel	Zurich Zoo
2004–2005	Ed McAlister	Adelaide Zoo
2006–2007	Karen Sausman	The Living Desert
2008–2009	Gordon McGregor Reid	Chester Zoo
2010–2011	Mark Penning	uShaka Sea World Durban
2012–2013	Jörg Junhold	Leipzig Zoo

Council

Since 1946 and still today, the WAZA Council is entrusted with the general direction and operation of the organisation. The Council also forms permanent committees as well as other committees and working groups. It appoints its members for the purpose of addressing specific issues and to satisfy WAZA's objectives. The first IUDZG Council consisted of only one person in 1935, the Association's first President Kurt Priemel, who was elected by his 17 peers to direct the informal gatherings of a selection of Europe's zoo directors. In 1947, the re-established Union expanded its Council, which it called the Board of the Union, to include a President and a Secretary who were elected by a majority vote of the Union's members. Both the President and the Secretary held office for three years and were eligible for re-election at the expiration of that term.

The original structure of the Board of the Union limited the power of decision-making to a small minority of the group. Increasing commitments and a growing membership led to the Board of the Union expanding in 1962 to include a President, First and Second Vice-Presidents, a Secretary, a Treasurer and one elected member, each of whom were in office for three years. By 1981, the Board of the Union was called the Council and consisted of a President, a Vice-President, a First Secretary and a Second Secretary. As of 1991, Council members held office for a term of two years instead of three.

Today, the decision-making power of WAZA rests in the hands of more individuals, reflecting how the organisation on the whole has grown to be more inclusive and member-driven over time. The officers and the Council members represent geographic regions according to a composition determined by Council and each member of the Council holds office for a period of two years. The WAZA Council consists of the President, a President-elect (who functions as the Vice-President)

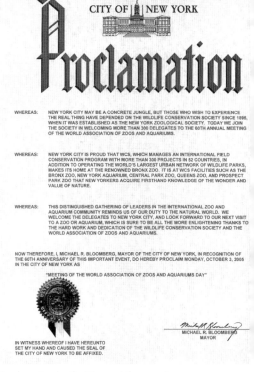

Proclamation by the Mayor at the occasion of the 2005 Annual Conference in New York

Participants at the 1975 Annual Conference in Colorado Springs (© WAZA)

and seven other members. The handover of the presidency occurs at the Annual Conference of the preceding year (for example, Mark Penning took office as WAZA President in October 2009, and his term lasted until October 2011). Since 2009, four members representing regional associations have been invited to participate in Council meetings, serving to add their valuable perspectives and insights to the group.

Annual Conferences

Each year since 1946, WAZA has held an Annual Conference. From the very beginning the purpose of the conferences has been for members to share zoological information, to promote cooperation, to accept the reports of committees and working groups, and to approve WAZA's financial statements and budget. The conferences have always included both an Administrative Session and a Scientific Session, and have often been filled with lively, sometimes contentious discussions on a vast array of zoo- and aquarium-related issues. Appendix IV lists where conferences have taken place.

National and Regional Associations

Since their inclusion into the WAZA membership in 1991, recognised national and regional associations have increased in number from eight to 24. Today, they represent a broad geographical range and serve as valuable portals, maximising the extent of WAZA's international reach and influence.

National and regional associations
that are members of WAZA as of December 2011

Acronym	Current Name of the Association
ACOPAZOA	Colombian Association of Zoo Parks and Aquariums
AFDPZ	Association Française des Parcs Zoologiques
AIZA	Iberian Association of Zoos and Aquaria
ALPZA	Latin American Association of Zoological Parks and Aquariums
AMACZOOA	Mesoamerican and Caribbean Association of Zoos and Aquariums
AZA	Association of Zoos and Aquariums
AZCARM	Mexican Association of Zoos and Aquariums
BIAZA	British and Irish Association of Zoos and Aquariums
CAZA	Canadian Association of Zoos and Aquariums
DAZA	Danish Association of Zoos and Aquaria
DTG	Deutsche Tierpark-Gesellschaft
DWV	Deutscher-Wildgehege-Verband
EARAZA	Eurasian Regional Association of Zoos and Aquariums
EAZA	European Association of Zoos and Aquaria
JAZA	Japanese Association of Zoos and Aquariums
PAAZAB	African Association of Zoos and Aquaria
SAZARC	South Asian Zoo Association for Regional Cooperation
SAZA	Swedish Association of Zoological Parks and Aquaria
SEAZA	South East Asian Zoo Association
SNDPZ	Société Nationale des Parcs Zoologiques
UCSZ	Union of Czech and Slovak Zoological Gardens
UIZA	Italian Union of Zoos and Aquaria
VDZ	Verband Deutscher Zoodirektoren
ZAA	Zoo and Aquarium Association Australasia

The Relationship between WAZA and Regional Associations

By Sally Walker, Chair of the WAZA Associations Committee and Founder/Director of the South Asian Zoo Association for Regional Cooperation (SAZARC)

Why Associations Came into Being

The first and oldest zoo association in the world is still operative. The Verband Deutscher Zoodirektoren (VDZ) was established when zoo directors attending the Antwerp Zoo animal auctions learned how agreeable and useful it was to share experiences related to their profession. The American Association of Zoological Parks and Aquariums (AAZPA, now AZA) was established in 1924. There was a 15-year gap between its establishment and the Japanese Association of Zoological Gardens and Aquariums (JAZGA, now JAZA) in 1939; and 27 years between JAZGA and the Federation of Zoological Gardens of Great Britain and Ireland (now BIAZA), founded in 1966.

After this time there was a sort of quantum explosion of national and regional zoo-association births, such as the founding of the European Association of Zoos and Aquaria (EAZA) and the Australasian Regional Association of Zoological Parks and Aquaria (ARAZPA, now ZAA), so that by 1990 there were 18 and by 2011 there were 37 associations, 24 of which are members of WAZA.

Reasons for creating zoo associations varied from desperation, as in the case of war exigencies, to enthusiasm to improve and become part of a global community. In the case of VDZ (and for that matter IUDZG), members sought companionship and a sharing of experience and expertise.

AAZPA originally started as an affiliate of a related association but after several years broke off on its own to pursue an identity as a scientific institution, in order to address the wildlife conservation issues that had emerged in the early 1970s. JAZGA was a forced birth for survival because finding sustainable food resources for animals, protecting them, dealing with escapes in the chaos of war and acquiring animals had become possible only through bonding and cooperation. The Federation of Zoological Gardens of Great Britain and Ireland was established to set standards of animal welfare and husbandry at a time when many new zoos were being opened, the operation of which was often less than satisfactory, affecting the reputation of the better zoos. Later associations were established as variations on these major themes.

WAZA and National and Regional Associations

In 1986, IUDZG appointed Frederic Daman, Director of the Antwerp Zoo, to be its Zoo Liaison Officer, with a specific mandate to promote closer collaboration between IUDZG, CBSG, ISIS, IZE, and the regional and national associations. This was not an easy task because it was the first time that the subject had been so openly discussed within IUDZG. Two paths were taken. First, to use the strong drive for development that characterised the regional associations, giving them a forum to demonstrate their progress. Second, to propagate collaboration directly with IUDZG or in cooperation with ISIS and CBSG.

Participants at the International Zoo Associations meeting in Front Royal in 1990 (© Sally Walker)

Until the 1990s, there had been no event specifically to gather the (then) 23 regional and national zoo associations together. The AAZPA Board, whose members were acquainted with some zoo associations from projects and training, took up this initiative. Invitees to an International Zoo Associations meeting were heads of zoo associations around the world and influential individuals who were potentially useful in promoting and sustaining global communication and cooperation among all of the associations.

On 23–25 April 1990, 41 representatives from zoo associations, zoos and other appropriate organisations from several continents gathered at the Front Royal Conservation Centre. Roger Wheater, Director of the Edinburgh Zoo, represented IUDZG as its President. Leobert de Boer, Director of the National Foundation for Research in Zoological Gardens in the Netherlands, gave a seminal presentation on the need for an international communications set-up for all zoos and how it might be organised. There were presentations from all associations, and it was a stimulating and moving event with an atmosphere of progress and change.

One of the main conclusions of the International Zoo Associations meeting was that a global zoo network should be established for the exchange of information between all zoo associations. There was a contingent that seemed to want a separate communications network that would include IUDZG, CBSG, ISIS, and the regional and national associations. IUDZG, however, had almost completed its struggle with its own membership policy and was close to having a workable outcome. This would result in the extensive constitutional changes that permitted recognised regional and national zoo associations to become association members of IUDZG. No one then knew that this would definitely come about, but IUDZG continued its role as liaison. The IUDZG President was asked to send a letter to all regional and national zoo associations laying out an approach intended to improve international communication among zoo associations.

The approach was simple; for example, each regional and national association should circulate their own association information among all other associations, including IUDZG, ISIS and CBSG, and from other associations to their own members. Frederic Daman, already IUDZG liaison, was to monitor the process and also seek a representative from each association. Little did they know that an opportunity would emerge in a little over a year, when two seminal events would occur: IUDZG would welcome regional and national associations into the Union as members, and the Internet would become popularised, providing extensive opportunities for communication.

In 1991, IUDZG changed its Constitution to include regional and national associations as potential members. The landmark decision resulted in a change in the character and function of IUDZG, broadening its influence and making it more truly representative of the world's zoos. By 1992, there were five association members, another five joined in 1993 and at the time of writing there are 37 known zoo associations representing a great part of the world, 24 of which are members of WAZA.

Successes and Constraints of Associations

Association heads regularly meet at WAZA's Annual Conferences in order to discuss subjects of mutual concern among associations and to communicate with WAZA. The meetings are usually well attended and topics such as matters arising from the Committee for Population Management (CPM), new global projects and problems are discussed. Each region is encouraged to give a report of their activities to WAZA plenary and to participate in committee meetings, including their own Associations Committee. Language differences between countries have been a constraint for zoo associations, particularly in international and some regional training sessions, meetings and conferences.

Robert Wagner, Roger Wheater and Kamal Naidu (from left) at the International Zoo Associations meeting in Front Royal in 1990 (© Sally Walker)

Another constraint to regional associations in low-income countries is their inability to bear the high cost of membership, flights and other expenses incurred by attending WAZA Annual Conferences, and technical workshops and training sessions. Ways to deal with these problems are regularly discussed although are not yet resolved.

The overarching success of the zoo associations is their development over the years to lead member zoos in embracing and striving for improved conservation breeding, research, education, training, technical guidelines, ethical behaviour and animal welfare. Over the last 25 years, WAZA has made dramatic changes in its relationship with the associations and, in so doing, has improved the potential of the associations to work together internationally on the most important activities of zoos.

The Changing Face of WAZA

I hope that all the decisions of great moment for this organisation will always be given proper consideration, that we will not act in haste in order to regret at leisure, but also that we will never again take so much time to make a decision of such importance.

Roger Wheater,
IUDZG President 1989–1991

IUDZG began as an 'old boys club', which carefully hand-picked its members by scrutinising each candidate's status and qualifications before admitting them. In the early days, the members enacted traditions resembling what might be found in a fraternity. New members received a tie to signal their unity with the group, members politely referred to each other as 'Colleague' and each time a member died the group performed a ritual of standing for a minute of silence at the beginning of an Annual Conference's Administrative Session to pay their respects. IUDZG members went so far as to write a special resolution for their esteemed colleague Axel Reventlow, Director of the Copenhagen Zoo, which was unanimously adopted by the membership. It read:

'Be it resolved at this first meeting since his departure to his eternal rest, that in token and respect for our beloved colleague, this Union unanimously expresses its sorrow at his passing and its continuing affection for his memory; and be it further resolved that [this] resolution be spread upon the Minutes of this Meeting and that a copy thereof be sent to [his wife] as evidence of the sincere sympathy of his colleagues.'

Participants at the 1962 Annual Conference in San Diego (© WAZA)

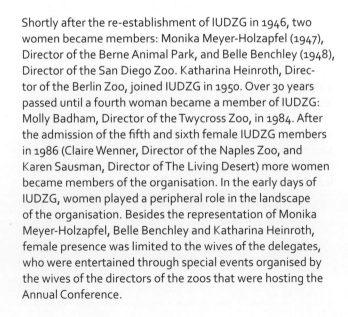

Shortly after the re-establishment of IUDZG in 1946, two women became members: Monika Meyer-Holzapfel (1947), Director of the Berne Animal Park, and Belle Benchley (1948), Director of the San Diego Zoo. Katharina Heinroth, Director of the Berlin Zoo, joined IUDZG in 1950. Over 30 years passed until a fourth woman became a member of IUDZG: Molly Badham, Director of the Twycross Zoo, in 1984. After the admission of the fifth and sixth female IUDZG members in 1986 (Claire Wenner, Director of the Naples Zoo, and Karen Sausman, Director of The Living Desert) more women became members of the organisation. In the early days of IUDZG, women played a peripheral role in the landscape of the organisation. Besides the representation of Monika Meyer-Holzapfel, Belle Benchley and Katharina Heinroth, female presence was limited to the wives of the delegates, who were entertained through special events organised by the wives of the directors of the zoos that were hosting the Annual Conference.

On Being a Woman in an 'Old Boys Club'

By Karen Sausman, WAZA President 2006–2007 and Past Director of The Living Desert

At my very first AAZPA meeting in 1967, I stepped up to the registration table to pay my fees and I was sweetly asked whether I was 'a lady or delegate'. I had to make a quick decision and decided on 'delegate'. I hoped I could still quietly be a 'lady' as well. My approach then and still today is that I am a zoo professional first and foremost, and my colleagues soon came to see me as just that. When I attended my first IUDZG meeting as an observer in 1980, I adopted the same approach and within days I felt accepted. When I became a member in 1986, most of the colleagues had come to know me as just another zoo director dedicated to furthering our profession. By the time I was elected to Council I had been around long enough to qualify as being a member of the old boys club, so I would like to think my election as President was easily accepted. I certainly felt warmly supported by everyone.

I have always felt that race, gender or any other label does not define the person. Thus, I tried hard to never use my gender as an excuse or advantage. I wanted to be the best zoo professional I could be and to support our organisation by taking on any task I felt qualified to complete. I was very fortunate to be accepted by colleagues around the world who supported my efforts. I owe a special debt of gratitude to my colleagues for sharing their time and wisdom with me, and the faith they had in me in my many, many projects over the years.

The list of people who have encouraged me from my early years in the 1960s and 1970s would fill a book. Lester Fisher was perhaps the very first to see that special spark that I had for the world of zoos when he was the Director at the Lincoln Park Zoo, and I started as a volunteer in the animal nursery and reptile building. He was soon followed quickly by William Wooden who encouraged me to follow my dreams and start The Living Desert. Gerald Durrell set another seed about what a zoological institution could and should be. And my early colleagues who, for whatever reason, felt inclined to point me in the right direction and give me many helping hands: William Conway, George Rabb, Chuck Bieler, Charles Hoessle, Ulysses Seal and James Dolan. Clayton Freiheit, in particular, introduced me to the international zoo world virtually demanding that I get involved.

Participants at the 1969 Annual Conference in New York (© WAZA)

This led me to a whole new group of friends and colleagues that absolutely affected my view of what zoos can do collaboratively around the world and how important our mission was. People such as Roger Wheater, Jeremy Mallinson and Gordon McGregor Reid, and all my colleagues who encouraged me to join WAZA's Council and ultimately to become its President.

I often joke that because in the 1960s there were no woman zoo directors, the only way to become one was start your own zoo! I had worked at the Lincoln Park Zoo while going to college and found myself reading the early Gerald Durrell books and thinking about his dream of building a zoo that was founded on conservation issues and breeding endangered species. When I found myself with the opportunity to help develop a local nature centre, I quietly decided I would provide the 'nature centre' and then move on to making it a true conservation and education centre for the world's deserts.

At the time the local population of our area was tiny and we were surrounded by miles of open natural desert. I explained our mission of 'saving the desert' and was usually asked 'from what?'. Thus, our greatest challenge was in creating an awareness of the beauty and complexity of the desert ecosystem. Years of effort by many wonderful supporters, staff and colleagues have made the effort a modest success.

Just as it has taken many years to incorporate more women into IUDZG membership, the pace of change and improvement to IUDZG's organisational structure has also taken time. In the early days from its re-establishment in 1946 to the 1970s, IUDZG was like a social club, with members meeting once a year in the convivial surroundings of different zoological gardens, networking, exchanging ideas and resources and socialising. IUDZG's small constituency at the time had restricted lobbying power and no significant financial base. Although members were vigorously creating resolutions and contributing as best they could to the cause of conservation, the organisation was limited in its influence.

Participants at the 1986 Annual Conference in Wroclaw (© Sally Walker)

It was not until the 1980s when the urgency of the issues facing wildlife and a growing understanding of the complexity of conservation began to force the organisation to look inwards in order to better position itself to affect more substantial change. Discussions about the expansion of membership to include institutions instead of just individuals, and of inviting national and regional associations to join, were frequent during the 1980s. However, there was often a lack of consensus among members, repeatedly forcing important decisions related to organisational change to be postponed until future meetings.

By 1990, there was a pervasive awareness by members that they either had to 'change or go extinct'. In other words, if they did not enact significant organisational change to grow the constituency and influence, IUDZG would cease to exist. Alarmed by this conclusion, members took the bull by the horns and ushered the organisation into 'the turbulent nineties'.

The Turbulent Nineties

By Willie Labuschagne, WAZA President 2000–2001
and Past Director of the National Zoological Gardens Pretoria

In the 1990s members of WZO prepared themselves with a certain degree of apprehension for the end of an era. The year 2000 was approaching and while most people and organisations were anxious to accept the new millennium with diligence and enthusiasm, WZO was still in a maelstrom of ambiguity, struggling to find itself.

The rebellion started in the mid-1990s when, after heated debates, which spanned more than one Annual Conference, we succeeded by majority vote in breaking away from the founding name 'International Union of Directors of Zoological Gardens'. To some, this name change was more than adequate and a sufficient enough sacrifice to disentangle ourselves from the criticism of being 'nothing more than an old boys club'. Great was the surprise when members continued to complain that the 'new' WZO still did not deliver! A mere change of name was simply not adequate.

These complaints had nothing to do with the leadership at that time and past presidents, the likes of Gunther Nogge, Peter Karsten, Roger Wheater (two of them Heini Hediger Award recipients) and others, did everything possible to ensure the survival of WZO and keep the organisation on track. But there was no 'track'. Members became very outspoken, and some even challenged and questioned their future association with the organisation.

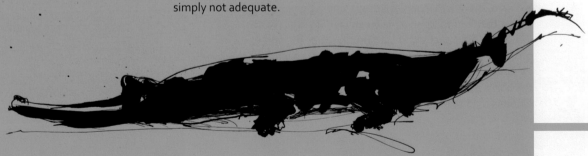

The 1999 Annual Conference in Pretoria was on hand. It was, therefore, no surprise that the theme for that Conference was 'Change or Die!'. Never will I forget how delegates at Bakubung scrambled under the African sun to find a solution for the future of WZO, an organisation which was deeply cherished. On the closing day members were still undecided and, as a solution, they unanimously delegated and empowered the incoming Council to create a new organisation to be introduced at the following year's Annual Conference in Palm Desert. That was the mandate and mission, which Council accepted with uncertain honour and apprehension.

IUDZG was originally founded in the mid-1930s. At that time, it was a European-based organisation with a very strong German-speaking influence. With the passage of time, zoos from the USA joined and shortly thereafter the IUDZG Council realised that they had something good and that it should be protected. Bylaws (a Constitution) were promulgated and membership to this prestigious organisation was by invitation only and was limited to zoos of outstanding quality.

These principles cannot be faulted. However, early on the organisation did not really expand and mostly represented members from a relatively small area of continental Europe, UK, North America, only one member from Africa and a few other isolated regions. Part and parcel of Annual Conferences was simultaneous translation in German and English, which was indicative of the regions represented at IUDZG meetings. It was, therefore, to be predicted that one of the first criticisms in the early 1980s was: 'Are we really a *world* organisation?'.

This resulted in a growing animosity amongst some members who zealously protected the character of the founding principles. But the future was clear, IUDZG had to grow and become truly international. This had to be accomplished by not sacrificing the basic principles of IUDZG and, moreover, without offending the founder institutions, some of whom already threatened to terminate their membership and even considered forming break-away organisations.

This more than anything else was the underlying concern expressed by the 1999 Council members in designing a new organisation – protect the interests of all.

A forerunner to the revolution was the need to have a secretariat. A proposal was accepted by both parties that ISIS could fulfil this function. The responsibility of ISIS was primarily to take minutes at meetings and to maintain the membership database. Although small in magnitude, this additional responsibility taxed the already limited manpower of ISIS personnel even more and it was realised that this arrangement should be discontinued. In fact, the annual cost of this service was something WZO could hardly afford at that time.

A strong plea was then put forward at the 1998 Annual Conference in Nagoya to establish our own full-time secretariat. The only flaw in this proposal was the expectation that the newly appointed director had to generate their own salary, something I personally strongly objected to. It became obvious that if we were serious about improving the organisation, we should throw everything overboard and start from scratch. The 1999 Annual Conference took place and Council was primed achieve its goal of creating a new organisation.

The first action was to discontinue using ISIS as the secretariat and to accept the offer of the National Zoological Gardens in Pretoria to perform this function at a substantially reduced cost. I had the dubious honour to act as Honorary Director; which I later regretted as some members spread a defamatory rumour that I was preparing myself for the permanent position! However, I soon realised that if we wanted to succeed, we should be prepared to ignore feeble attempts by the minority to derail us.

WZO Council consisted of Bernard Harrison, Ed McAlister, Stephen Wylie, Alex Rübel, William Dennler and myself. Our first meeting took place at the National Zoo's Game Breeding Centre in Lichtenburg, South Africa. An excellent hideout for an uninterrupted three-day session. Inspired by the confidence entrusted by the members, the deliberations were started by critically dissecting the organisation from its roots and building on the expectations for the future. We soon realised that the existing Bylaws were a major stumbling block and were not representative of those expectations for the future. Therefore, the existing Bylaws had to go and a complete new set had to be written, and this would imply adopting a new name and a new logo. We expected that we could still get away with new Bylaws and a new name, but the logo was precious and would require extremely delicate lobbying and negotiations.

The act of suggesting a new logo could destroy the only remaining bastion of the older generation of IUDZG, which so passionately protected the original logo. The zebra (or was it a quagga?) had to undergo a complete revision or go! The international marketing agency Young and Rubicam was employed to come on board and adopt WZO as a pro bona client. Their mission was to design a new logo which had to incorporate remnants of the old logo.

Critical to the new Bylaws was a clear identification of membership and, more importantly, a democratic process of electing Councillors, in particular the President. The members would expect much more than just new Bylaws, a new name and a new logo. Dramatic as it might be, this would not address the underlying problems and might be considered as an extremely superficial effort by the Council in trying to improve the organisation. It was essential that the mistakes of the past were not repeated and, in order to get the organisation moving, we required tools to accomplish this.

It was imperative for the future that a full-time secretariat was established and a full-time director appointed. We were adamant that the director should not be expected to generate their own salary and, for that reason, a five-year business plan had to be developed to accomplish the goals. It was also agreed that a newsletter would be published to keep the members informed of Council's progress.

However, instead of doing good, the newsletter almost caused irrevocable damage. Members became aware of our intended plan to appoint a director and one region immediately offered the name of a candidate for the position. In principle there was nothing wrong with this, but when Council was informed that only this candidate and no one else would be accepted by that region, all such negotiations were immediately withdrawn and the recruitment was placed on hold indefinitely.

And all of a sudden Council was at a crossroads. We required strong diplomacy if we wanted to succeed. Issues such as the selection of a director might jeopardise the whole plan. Therefore, we concentrated on the Bylaws, name, logo and business plan. In fact, they were enough to carry us through. A further meeting was arranged prior to the Annual Conference in order to have well-polished proposals ready for the event.

The next meeting was attended by the Young and Rubicam artists and Council viewed with great sympathy their ingenious methods of trying to save the original logo. It was just not possible and it was agreed that the logo should go. Young and Rubicam received a new mandate and produced a stunning logo depicting the earth, the air and the ocean – yet to be approved. The proposed logo was published in the newsletter and members were invited to comment. It was enlightening to experience the response.

Instead of a downright rejection, zoo directors, many of whom were 'internationally renowned logo designers', made the most profound suggestions, such as moving the eye of the bird a millimetre to the left, moving the tail of the fish a little bit to the right, the eye of the elephant should be a fraction larger, etc. This was a breakthrough. Instead of being dismissed there were many admirers just wanting to add their own little bit. Of course all of the proposals were 'incorporated' and everyone was happy.

The 2000 Annual Conference in Palm Desert arrived and the agenda for the event was carefully planned. I was to 'soften' the constituency by reminding members during my opening speech of our mandate received during the 1999 Annual Conference and give a brief summary of the areas Council would be concentrating on. I also had to assure the members that all of our proposals would be going through a proper election process.

A detailed report was presented during a session called 'Profile for the Future'. Members were informed of the need to establish a full-time secretariat, to appoint a permanent director and to find a suitable location for the proposed office. The session was followed by a presentation of the proposed business plan, which would help to make all of this possible. A brief opportunity was offered for discussion. We opened the floodgates when we announced our proposals for the new name, new logo and, more importantly, the new Bylaws.

A senior zoo director responded that Council might have received a mandate to recommend changes to the Bylaws, but never to change the entire organisation. The response stunned us. We never expected such a blatant attack! Fortunately, and before any of the Councillors (who were all seated at the main table) responded, another equally senior zoo director announced that you cannot go halfway if you have to remodel an organisation.

Needless to say, Councillors did not participate in the afternoon/evening visit to the San Diego Wild Animal Park. We collapsed in my spacious suite (courtesy of Karen Sausman) and carefully deliberated on the reaction by the members. We convinced ourselves that the criticism earlier that afternoon was a 'flash in the pan' and not representative of the entire constituency. The following morning members would be allowed to comment on our proposals – no election yet, only dialogue.

The following day, during coffee and lunch breaks, Councillors were faced with one-on-one interrogations and resistance by prominent members representing major constituencies. Helmut Pechlaner (a respected colleague and a good friend of mine for many years) represented one such group. He was obviously a leading mediator and careful negotiations had to be implemented.

I vividly recall when Helmut said: 'Willie, we will give you the name, but not the logo'; later that morning: 'Willie, we will give you the logo, but not the Bylaws'. And so it carried on for four gruelling days. The biggest task was to convince the protagonists they were not losing their organisation, but that we were simply trying to improve it to meet the challenges of the future. Although we were confident that we had the majority support, it was still important that the minority should be given the opportunity to express their reservations. Even if this round was won, in order to take the process forward it was important that everyone gave their support.

We waited until the eleventh hour to put our recommendations to the test and invited members to cast their votes. Each case was handled separately. In other words, members could vote for or against each of the main proposals separately and some did. But when votes were counted, supervised by independent referees, the results turned out to be overwhelming support in favour of our recommendations.

The World Association of Zoos and Aquariums (WAZA) was born with a sexy logo, a new name, impressive Bylaws, a dramatic five-year business plan and a mandate to proceed with the establishment of a full-time secretariat.

This was not the end but the beginning of a new era. Council now had all of the support and means to set the ball rolling. The success of the five-year business plan and the opportunity to establish a full-time secretariat depended exclusively on a very aggressive membership drive. A certain membership growth per annum had to be achieved in order to meet the goals. We unequivocally believed in our ability to meet the challenge and were prepared to cross the Rubicon. The campaign to recruit a director was started. Oblivious of the previous threats, the position of WAZA's first director was advertised internationally. Moreover, all regions were invited to put forward offers to house the secretariat.

A three-man commission, consisting of Alex Rübel, Bernard Harrison and myself, was charged with the responsibility of interviewing candidates and making the final selection of both the candidate and the location of the secretariat. Peter Dollinger took office as WAZA's first Executive Director in Berne, Switzerland in October 2001 and the rest is history.

The work of Council was not completed. Our commitment to a greater democracy still had to be addressed, committees in-line with our new Bylaws had to be appointed and conservation hot spots had to be identified. Since its inception, there has never really been a democratic process of electing members to Council. This was something close to my heart and I was determined to address this in a most exact way. The 'three basket nomination programme' was adopted by Council and the 2001 Annual Conference in Perth was approached with the knowledge that our job was done. The mandate received from our members two years earlier was accomplished. The mission was completed.

Sometimes I miss the small group of zoo directors of yesteryear. The black-and-white group photographs, the quibbling over simultaneous translations, the tie new members received as a symbol of patriotism and our most distinguished manner of addressing one another; always saying 'Good morning Colleague', 'how are you Colleague', 'see you next year Colleague'.

The zebra is dead. Long live WAZA!

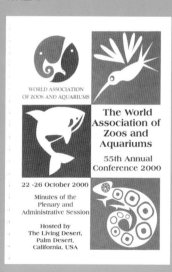

Cover of the minutes of the Plenary and Administrative Session of the 2000 Annual Conference in Palm Desert, featuring for the first time the new name and logo

The momentum originating from the groundbreaking organisational changes that took place in 1991 continued for the remainder of the 1990s and into the new millennium. Further big organisational changes were adopted, such as three name changes (IUDZG – WZO in 1992, WZO in 1998 and WAZA in 2000), two new logos and five revisions of the Bylaws (Constitution). Today's WAZA is a more malleable organisation, which is responsive and adaptable to change. Although it is still closely bound to the core principles of its founding members, it has come a long way from its social-club roots.

WORLD ASSOCIATION
OF ZOOS ANS AQUARIUMS

The Changing WAZA Logo

The WAZA logo has transformed over time to reflect the organisation's evolving priorities in a changing world. The first logo featured a zebra with a picture of a directional and was used from 1956 until 1999. The second version of the WAZA logo debuted in 2000. It was a circular design depicting an elephant, a bird and a fish in unison. The three animals represented the animals of the earth, sea and air, and were designed to illustrate the symbiosis between the three vital elements. The third and current version of the WAZA logo replaced the earlier version in 2008. Its design was modified to have better visibility and a more modern look. The slogan 'United for Conservation' became an integral part of the logo as did the full and abbreviated name of the organisation: WAZA.

World Association of
Zoos and Aquariums
WAZA | *United for
Conservation*®

IUDZG/WAZA logos over time

Marketing Initiatives

The *Zoo Future 2005* report set out four principal priorities for zoos for the period between 1995 and 2005. They included: (1) building cooperative linkages between *in situ* and *ex situ* conservation programmes; (2) improving animal well-being in zoos; (3) building a worldwide network; and (4) improving public perception, understanding and appreciation of zoo missions. The WZO Marketing/Public Relations/Education Committee was formed in 1995 in order to help refine, develop and implement the fourth priority.

Sherry Sheng from the Metro Washington Park Zoo served as the first Chair of the newly formed Committee, establishing four working groups each with a separate focus: public perception, message development, communications and the education/IZE liaison. The communications working group agreed that, in the short term, a marketing initiative should be developed, which would incorporate WZO philosophies including its mission, vision and values. The group also agreed that the initiative should focus on the principle of conservation for future generations.

The first international zoo marketing conference led by WZO was held in Aalborg in 1997. The theme of the event was 'Marketing Zoos Beyond 2000' and spanned over three days, which were filled with discussions, workshops and networking opportunities for 140 people representing 70 zoos worldwide. In 1999, the Amsterdam Zoo sponsored a zoo marketing conference for WZO and EAZA focusing on determining what 'zoo visitors want' and on 'gaining support'. In 2001, a further marketing conference took place in Tenerife, highlighting 'Conservation through Commerce'. By 2002, WAZA's website went live, extending the reach of the organisation's marketing arm further into the conservation community. In addition to the proceedings of the Annual Conferences, WAZA regularly publishes the WAZA News and the WAZA Magazine.

Cover of the proceedings of the 1956 Annual Conference in Chicago, featuring the original IUDZG logo for the first time

Homepage of WAZA website in 2011

Advertisement for the marketing conference in Granby in 2011
A man-made totem as 'marketing action' in Granby in 2011 (© WAZA)

In 2003 and 2005 the fourth and fifth international conferences on zoo marketing and public relations were held in Budapest and Münster, focusing on 'Marketing Zoos and Conservation through Campaigns' and 'The Colour of Zoo Marketing is Green', respectively. The next conference took place in Pretoria in 2007, centering on the theme of 'Challenges of Zoo Marketing from Basics up to Global Conservation Campaigns'. In 2009, a full-time position of WAZA Marketing and Communications Officer was created, enhancing the professional operations of WAZA marketing and enlarging the spectrum of the organisation's marketing and public-relations capacity. In 2011, the seventh zoo marketing conference was held in Granby, focusing on 'Strategic Marketing in Action'.

Marketing for the Global Zoo Community

By Jörg Junhold, Chair of the WAZA Marketing Committee and Director of the Leipzig Zoo

In March 2002, the WAZA Marketing Committee was established as the successor of an informal marketing committee that had been working since 1996. The new Committee was a result of the follow up of the WZO/CBSG Workshop 'Zoo Future 2005', which took place in 1995. The WAZA Marketing Committee mainly consists of experienced zoo directors and serves as a source of expertise in order to help develop the 'WAZA brand'. It guarantees that WAZA will benefit from the experiences of zoos all over the world.

In order to generate public interest in zoos, marketing is of greater importance to WAZA today than ever before. Marketing is more than just advertising; it entails planning and implementing all of the activities of an organisation in order to satisfy the wishes and needs of the customers. But who are the 'customers' of WAZA, only members and zoo experts or the broader public as well? And how do we know which direction to move forward strategically?

Dating back to 2006, the WAZA Executive Office together with the Leipzig Zoo carried out a marketing survey involving 252 members and regional institutions (of which 125 responded). A detailed profile was generated of the participating zoos, which included visitor numbers, budgets, species kept and the marketing tools used. Three-quarters of the respondents ranked the importance of marketing issues to be very high and were willing to support WAZA's marketing efforts. Over 90% of the participating zoos indicated that they were interested in having WAZA facilitate worldwide cooperation as well as lobby for and support conservation work. A total of 85% of respondents believed that WAZA should function as an umbrella institution for all zoo organisations and 90% believed that WAZA should serve as a think tank for strategic issues as well.

One of the basic questions that I ask myself is whether WAZA has the potential to become a publicly known environmental organisation, which each single member zoo can benefit from. This will need improved professionalism in the field of marketing and public relations, which is already taking place.

The first International Marketing Conference in Aalborg dates back to 1997. Many presentations highlighted the effectiveness of new marketing tools employed by zoos, which were making increased use of website services, social networks, TV series, twitter, radio spots and many other measures. These marketing tools can also be used by WAZA.

One visible step towards intensified professional work is the new position of WAZA Marketing and Communications Officer in the Executive Office since 2009. This has already led to a better use of member zoos' press releases, for example on the new WAZA website.

The bonds with international partners such as the Convention on Biological Diversity (CBD), IUCN and the Convention on Migratory Species (CMS) will, on the one hand, help WAZA to become more integrated into international conservation work and, on the other hand, will improve WAZA's reputation as a serious player in the conservation field.

I believe that WAZA members are not yet fully aware of their growing role and responsibility for conservation, education and research, nor do they recognise the public influence that already exists. Each of us needs to reflect on the fact that in addition to marketing our individual zoos, we should also highlight the united organisation behind it – WAZA.

Chapter 3

The World Association of Zoos and Aquariums' Commitment to Conservation

As is the case with all organisations who have a long and rich history, IUCN and WAZA have had to reform and reinvent themselves along the way to become the truly modern global conservation networks that they are today. In doing so, both have built on the simple but powerful idea: by working together, we can achieve more.

Julia Marton-Lefèvre, IUCN Director General

European bison (© Martin Wehrle)

WAZA's Evolution into a Conservation Organisation

The 1930s to the 1950s

Although the economic challenges in the early 1930s were of overwhelming importance and, therefore, thoroughly discussed at the earliest formalised zoo-director meetings, nature conservation has always featured on their agenda. In 1932 and 1933, for example, the meetings in Berlin and Frankfurt took place alongside the meeting of the International Society for the Protection of the European Bison. Furthermore, at the International Association of Directors of Zoological Gardens founding meeting in Basel in 1935, it was made clear that in addition to exchanging important information between zoos, nature conservation should always feature on each meeting's agenda. Another example of the early Association's commitment to nature conservation was evidenced by the fact that at the 1936 meeting in Cologne, the state representative for nature conservation was invited to give a talk during a special session, normally reserved for the German zoo directors.

IUDZG members from the very first meeting of the revived organisation in 1946 expressed their concern about the protection of wildlife and wild places. In 1947, a role for zoos in conservation was drawn up and members were asked to observe the following principles:

Participants at the 1959 Annual Conference in Copenhagen (© WAZA)

- Zoos should abstain from dealing in protected species
- Zoos should try to ensure the reproduction of species in danger of extinction
- A list of rare and protected species should be drawn up
- An organisation should be set up to manage breeding centres for these species

A second conservation milestone for IUDZG was to become a member of the International Union for the Protection of Nature (IUPN, now IUCN) in 1949. Joining this newly fledged international union signified IUDZG's early commitment to nature protection. Members were aware of the increasingly vulnerable state of the environment and understood that many animal species would be more and more threatened by extinction in the coming years. In 1952, IUDZG President Armand Sunier, Director of the Amsterdam Zoo, read a paper entitled 'What Does Our Union Stand For?' (Appendix IV), which highlighted this outlook. He stated that:

Giant panda (© Christian Schmidt)

Letter written to the Chinese government on giant pandas in 1958 (© WAZA)

'Everywhere in the world, nature is inevitably becoming more deeply modified by the constant incursions of the human population. In the very near future it will certainly no longer be possible to preserve all the natural surroundings to a sufficient extent to assure the survival in complete liberty of all the species threatened by extinction.'

In addition to their concern for the environment and the threat of species extinctions, members in the 1950s were alarmed by the excessive trafficking of live animals, especially birds. In 1954, the 37 members of IUDZG acted on their concern by unanimously accepting a request from the International Council for Bird Preservation (now BirdLife International) calling for the restriction 'on the export and import trade of wild caught birds and only to permit the export of rare and protected species in small numbers to zoos and *bona fide* aviculturalists for scientific study and breeding'.

The year 1958 marked another milestone for IUDZG. It was the first time that the Union attempted to influence a government to enact conservation measures. IUDZG President Walter Van den bergh, Director of the Antwerp Zoo, wrote a letter to the President of the 'Chinese Popular Republic' asking for 'rigorous measures to be taken to protect the giant panda from being hunted or captured'.

Participants at the 1963 Annual Conference in Chester (© William Conway)

The 1960s

As in the rest of the industrialised world, the 1960s were a time of growth for IUDZG on the world environmental scene. The decade started with significant momentum. In 1960, IUDZG gave its first support, in the form of a resolution, to an international wildlife project, the Africa Special Project (1960–1963). The project was launched by IUCN in order 'to conserve the existing African National Parks and faunal preserves and the wildlife populations living on marginal and other lands in different parts of Africa'. The members of IUDZG wrote a resolution stating their 'full appreciation for and support of this project, with the fervent hope that the project launched will succeed in maintaining and preserving all African wildlife for the benefit and edification of the future generations for all mankind'.

Sumatran orangutan (© Christian Schmidt)

Monkey-eating eagle (© Christian Schmidt)

In 1963, following their earlier commitment to limit the trafficking of live birds, IUDZG extended their commitment to orangutans. It was the second time that members united behind self-imposed standards of control in respect to wild-caught animals. Members unanimously decided that 'all Union members should refuse to purchase orangutans which have been caught illegally and to offer to assist IUCN by keeping any orangutans and holding them at the disposal of IUCN until such animals could be resettled at a later date'.

Bernhard Grzimek, Director of the Frankfurt Zoo, presented a paper at the 1963 Annual Conference in Chester on the need for 'biophylaxis' or what is known today as wildlife management. He critically examined the state of African nature reserves and highlighted the fact that 'all fundamental knowledge' about the wild animals living in the reserves was absent: 'We do not know what age is reached by zebras, gnus, Thomson's gazelles, hyenas and leopards; we do not know how rapidly they multiply, what they live on, where they go in their seasonal wanderings and to what diseases they succumb to'. He stated that 'not until we know more about the habits and needs of wild animals shall we be able to define our aims with precision'. His paper was a call to action for greater scientific research on wild-animal species. It was pragmatic, 'the most successful and the cheapest way to conserve an animal species in its original form is to breed it in its habitat'; and forward thinking about the limits of wildlife management, 'however hard we fight and whatever sacrifices we may make, we shall only be able to reserve for our wild animals very small residuary natural areas'.

In 1965, IUDZG drafted another resolution related to limiting the trafficking of species, this time on the procurement and trade of monkey-eating eagles. The Union resolved that 'its members should not in the future purchase or trade in monkey-eating eagles since this bird is in grave danger of becoming extinct. The members of IUDZG will use their influence to discourage other zoos, animal dealers and collectors from capturing, exporting and trading monkey-eating eagles until such time as the wild population is secure'.

The year 1965 was also noteworthy because it marked the first time that a financial contribution was made by IUDZG to a conservation project. A total of £ 1000 was donated by the Union to a national park in Uganda towards the purchase of a light aircraft to be used in anti-poaching patrols and general park management.

The year 1967 was marked by a further resolution, one that both restated the decree for orangutans and monkey-eating eagles, and added three more species to an ever-growing list. In a style notable for its candour, it read: 'No member will purchase, offer to purchase, sell, offer to sell, capture, encourage the capture of, donate, accept as a gift, or deposit or trade the following species: Orangutan, monkey eating eagle, Zanzibar red colobus monkey, Galapagos tortoise, Aldabra tortoise'. As the list of species threatened with extinction increased, so did IUDZG resolve to be a part of the solution rather than to be part of the problem. By joining together under the banner of 'a resolution', Union members wanted to use their combined influence to discourage other zoos, animal dealers and collectors from capturing, exporting and trading the vulnerable species.

Bernhard Grzimek with okapi in 1974 (© Frankfurt Zoo)

Zanzibar red colobus monkey (© Christian Schmidt)

Participants at the 1965 Annual Conference in Berlin (© Berlin Zoo)

In 1968, in order to increase the strength of their contributions to conservation, IUDZG made a payment to IUCN's Survival Service Commission (SSC, now the Species Survival Commission). The SSC's principal role was to promote the preservation of endangered and threatened species of fauna and flora. It was also concerned with trends, which if not checked could cause a species to become endangered. IUDZG paid US$ 500 on top of the yearly IUCN membership fee to be directed towards the important work of this keystone IUCN Commission. Their hope was 'to have an influence in worldwide decisions about wild animals and wild places'.

Participants at the 1968 Annual Conference in Pretoria (© WAZA)

Aldabra tortoise (left) and Galapagos tortoise
(© Christian Schmidt)

WAZA and the IUCN Species Survival Commission (SSC): A Long History of Collaboration

By Simon Stuart, Chair of the IUCN Species Survival Commission

The close collaboration between the IUCN/SSC and WAZA goes back many years. The collaboration was already close when I started working with SSC in 1985. At that time, WAZA was IUDZG. The SSC had established a Captive Breeding Specialist Group (CBSG, now the Conservation Breeding Specialist Group) in the late 1970s under the leadership of the unforgettable Ulysses Seal, and the two bodies have always worked closely together. Both IUDZG and WAZA have been long-term and generous financial supporters of CBSG, and to this day the annual meetings of both bodies usually take place one after the other (CBSG first, then WAZA) in the same place. They have evolved in very complementary ways, with WAZA being the more formal gathering of the global leaders of the zoo community and CBSG being the flexible scientific network, providing much advice and support to encourage long-term growth in the zoo-community's involvement in conservation.

However, the interaction between WAZA and SSC goes beyond CBSG. When I started out, the SSC Steering Committee used to jointly approve applications for new international studbooks with IUDZG. Many of SSC's leading figures have been closely involved with WAZA. For example, George Rabb, Chair of SSC from 1989 to 1996, was the long-term President of the Chicago Zoological Society and he remains a pivotal figure in international species conservation to this day. Peter Scott, the charismatic SSC Chair from 1963 to 1980, was closely involved in the captive breeding of ducks, geese and swans, and worked closely with IUDZG. Russell Mittermeier, long-term Chair of the SSC Primate Specialist Group, is a frequent speaker at WAZA gatherings. And WAZA Past President Gordon McGregor Reid is Chair of the SSC Freshwater Fish Specialist Group. These are just a few names in what is a much longer list.

In recent years, the collaboration has deepened. WAZA has become a generous financial supporter of the SSC Chair's Office, both for myself and my predecessor, Holly Dublin. WAZA collaborated with CBSG and the SSC Amphibian Specialist Group to launch the Amphibian Ark, and plans are now afoot to collaborate on the first World Species Congress. Perhaps most symbolic of all, the WAZA Executive Office is now located at the IUCN World Headquarters in Gland, Switzerland.

The SSC and WAZA are highly complementary bodies with compatible missions. As the global conservation challenges increase, I look forward to our collaboration becoming much stronger.

Mixed colony of great white egret, spoonbill, grey heron and pygmy cormorant in the reed belt of Neusiedler See (© Erwin Nemeth)

The 1970s

The progress made towards conservation by IUDZG in the 1960s was not sustained in the 1970s. The most significant IUDZG contribution to conservation was a letter written in 1977 to the Chancellor of the Federal Republic of Austria urging that an internationally valuable ornithological region for migratory birds called Neusiedler See become a fully protected reserve. It was the second time that the Union attempted to influence a government to enact conservation measures.

WAZA and the Convention on International Trade in Endangered Species of Wild Fauna and Flora (CITES)

By Peter Dollinger, WAZA Executive Director 2001–2008

At its first meeting in 1946, IUDZG adopted a resolution calling upon governments to severely control the capture and trading of wild animals, and claimed an international agreement on this subject. This call was obviously premature. It was only in 1973 that the governments, based on preparatory work by IUCN, adopted the Convention on International Trade in Endangered Species of Wild Fauna and Flora (CITES), which entered into force on 1 July 1975.

In the meantime IUDZG adopted a series of resolutions relating to animal trade: in 1963, it was agreed that members should refuse to purchase any orangutans, which had been caught illegally. This was followed by another resolution in 1967, to the effect that no member will purchase, offer to purchase, sell, offer to sell, capture, encourage the capture of, donate, accept as a gift, or deposit or trade orangutans, monkey-eating eagles, Zanzibar red colobus monkeys, Galapagos tortoises and Aldabra tortoises unless the IUCN/SSC had given its consent. Two years later, Aldabra tortoises were removed from the list, and mountain gorillas and mountain tapirs were added.

In the early days of CITES, AAZ-PA was the only zoo association to attend CITES conferences. EAZA became involved from 2000 and WAZA in 2002, after the WAZA Executive Office had been established and Peter Dollinger, who previously had been the Head of the Swiss CITES delegation, had become WAZA's first Executive Director.

The first official appearance of WAZA in the CITES context occurred at the 18th meeting of the Animals Committee held in San José, Costa Rica, in April 2002. The Association introduced itself to the CITES community, offered to organise animal-transport training workshops and provide fact sheets as a follow-up to CITES Decision 11.102 on the relationship between *ex situ* production and *in situ* conservation.

Mountain gorilla (far left) and mountain tapir (© Christian Schmidt)

Subsequently, several animal-transport seminars were organised in collaboration with the late Peter Linhart of the Vienna Zoo and a set of 40 fact sheets, entitled *The WAZA network links ex situ breeding with in situ conservation*, was distributed at the 12th meeting of the Conference of the Parties to CITES (CITES CoP12) held in Santiago, Chile, in November 2002.

In Santiago, WAZA and the other zoo associations present agreed that priorities should be: to make WAZA better known and its mission better understood by the CITES Parties and major non-governmental organisations; to become an official partner organisation of the CITES Secretariat and of the International Air Transport Association (IATA) in the field of animal transport; and to work towards the facilitation of administrative procedures. The WAZA Executive Director used the opportunity to introduce WAZA as an expert organisation philosophically close to IUCN and explained the role that modern zoos have in conservation.

Subsequently, the CITES Secretariat asked the representatives of WAZA and IATA (Peter Dollinger, Eric Raemdonck and Peter Linhart, who was a member of the Austrian delegation) to prepare a draft decision of the Conference of the Parties. This was done and the Parties adopted the following text by consensus:

'The Secretariat shall, in consultation with the Animals Committee, liaise with the International Air Transport Association (IATA) and the World Association of Zoos and Aquariums (WAZA) with a view of concluding a Memorandum of Understanding in order to:

• Strengthen further collaboration in order to improve transport conditions of live animals;
• Establish an official training programme on animal transport;
• Facilitate the exchange of technical information relevant to animal transport between the Secretariat, the IATA Live Animals and Perishables Board, and the WAZA Executive Office.'

Following the Conference, drafting meetings were held at the CITES Headquarters in Geneva, Switzerland. Regrettably, IATA ultimately refused to sign the MoU. In spite of the setback, however, the two remaining organisations continued to cooperate closely, in particular after WAZA employed Thomas Althaus who until 2009 was Chair of the Animals Committee.

Zoo people at CITES CoP12 in Santiago in 2002 (© Peter Dollinger)

Participants at the 1981 Annual Conference in Washington, DC (© WAZA)

The 1980s

By the 1980s, IUDZG members had developed more sophisticated thoughts about the global scale and international context of conservation issues and zoo relationships, evidenced by the fact that there were more papers on conservation-related issues and more conservation-themed discussions at the Annual Conferences than ever before. Significant contributions to conservation initiatives started with the Union's decision in 1981 to donate US$ 1,000 to IUCN/SSC's CBSG. As with the earlier supplementary payment to IUCN's SSC in 1968, IUDZG hoped to forge closer links with IUCN/SSC's CBSG and thereby increase their influence on international decisions about wildlife. The Union's donation to CBSG was the first of many. The act of giving to CBSG carved out a place for IUDZG at the heart of the international effort to breed endangered animals. It also solidified an important relationship between two organisations that were both in a position to make significant contributions to wildlife conservation.

Participants at the 1982 Annual Conference in Rotterdam (© WAZA)

From the very start of the decade there was a general understanding by members that they 'must do more than just keep studbooks' in order to help avert the tide of mass extinction. Wilbert Neugebauer, Director of the Stuttgart Zoo, and Bart Lensink, Director of the Amsterdam Zoo, stated in a joint presentation given at the 1982 Annual Conference in Rotterdam that 'we will probably not achieve very much, because the zoos' potential is rather low and cooperation will be more or less casual'. Their assertion suggested that members were realistic about the scale of what they could accomplish. A substantial part of the 1982 Annual Conference was set aside to 'deal with the problems of conservation and cooperation'.

Lester Fisher, WAZA President 1984–1986
(© Lincoln Park Zoo)

IUDZG First Secretary Dick van Dam, Director of the Rotterdam Zoo, ignited the ensuing discussion in a paper called 'Conservation and Cooperation, an Introduction'. He stated that 'only when we have demonstrated that we are both ready and able to manage the populations of animals in our care, can we convince others that we are capable of serving the cause of conservation'. Parallel to this concept, he made the point that without involvement in conservation 'the continued existence of the zoo-world could not be guaranteed'. Dick van Dam's 'battle call' became the basis for what the Union would prioritise for the next 10 years.

In his first speech as new IUDZG President in 1983, Lester Fisher, Director of the Lincoln Park Zoo, stated that 'it is essential that we get ourselves well organised'. Encouraged by the previous year's discussions about conservation and cooperation, members were indeed getting themselves organised. To begin with, they came up with a list of 12 unanimously accepted policy statements, which underscored IUDZG's commitment to making species management through inter-zoo cooperation central to the organisation's contribution to conservation.

Ten of the Most Noteworthy Policy Statements

1. Coordination of Management Programmes
[IUDZG] RECOGNISES the increasing importance of scientifically developed programmes for the management of species. [IUDZG] RESOLVES that the Union requires its members to make every possible effort to cooperate in the development and coordination of such programmes.

2. Selection of Species
RECOGNISES the specific role of the Species Survival Commission of the IUCN to select species in particular need of captive-breeding programmes.
RESOLVES that the Union requires its members to support and assist where appropriate the efforts of the SSC. That the Union continues to develop its cooperation with IUCN's Captive Breeding Specialist Group and to make every effort through its members who are members of the group that the very serious responsibilities placed on it in respect of identification of species at risk and proposals for species management programmes are brought to the Union's attention.

3. Record System
RECOGNISES that a unified record system is essential to efficient species management. RECOMMENDS that ISIS be used where appropriate. Where for whatever reason, ISIS cannot be used, then it is further recommended by the Union that group or individual systems be adopted to be compatible with ISIS.

4. Species Management to take Precedence
RECOGNISES that the success of coordinated species management programmes will depend on key animals within the captive population being located within the correct collection. RESOLVES that the membership should be encouraged to consider as a first priority the movement of their animals on the basis of the needs of the species and not of individual collections.

5. Value and Ownership
RECOGNISES that the cost and to some extent the ownership of animals important in the context of the managed population may provide a constraint to the animal being placed into the collection where the best possible results can be achieved to the benefit of the species.
RESOLVES that members use their best endeavours to persuade their boards of management that the long term future of species both in the wild and in captivity will depend on successful breeding programmes. That such programmes can only be successful if constraints on animal movements are removed. One of these constraints is the cost of the animal and individual rather than group ownership and that every effort should be made to remove or minimise this constraint.

6. Studbook Keepers

CONFIRMS its role in ensuring that the most appropriate studbook keepers are selected.
RECOGNISES the important role of the studbook keeper where detailed scientific management plans have been developed and agreed.
RESOLVES that members who have studbook species will consider carefully the advice of studbook keepers in relation to the movement of animals where the species involved are the subject of a management programme and that members will not make moves or breed animals which would prejudice such a programme.
REAFFIRMS that members who are owners of endangered species be urged to register their animals. That transfers and other management moves should only be carried out after consultation with the studbook keeper.

7. Captive Breeding by Other than Zoos

RECOGNISES that there are other successful private breeders operating on a small or large scale with a limited number of species. That these operators often have considerable resources to support a specialised approach.
RESOLVES that members should encourage either as individuals or as members of larger management groups the involvement of appropriate private breeding groups within the species management programme.

8. Wider Membership

RECOGNISES that there is a majority feeling that the membership of the Union should be extended to include the Directors of other zoological institutions who conform to the membership qualifications and have the necessary stature.
AGREES that existing members consider within their own sphere of knowledge those institutions whose directors are not currently members of the Union, and who would qualify on both counts and bring this to the attention of Council.

9. National and Other Group Representation

RECOGNISES that in pursuing its leadership role it may be appropriate to develop more formal links with national bodies.
RESOLVES that (a) in the first instance relevant national bodies be asked to appoint as their representatives existing members of the Union; (b) invite national bodies to provide an observer as occasion demands and not as a right to attend all meetings of the Union.

10. Liaison with Conservationists

RECOGNISES that support of conservation organisations at local, national and international levels is vital to the long term success of species conservation.
RECOMMENDS to members that they contribute to develop the closest possible links with local and national conservation organisations.

By the middle of the 1980s, IUDZG was still encouraging efforts to collaborate and coordinate zoo activities on an international level. IUDZG President Lester Fisher's welcoming address in 1985 embodied the spirit of the time:

'In recent years this Union has discussed a wide range of matters relating to our policy in pursuit of our conservation goals. This dialogue has been most important and certainly there is little doubt that the conclusions that we have come to in developing greater understanding and making us better able to cooperate internationally in species management programmes, have enormous relevance at a time when the position of some species becomes of greater and greater concern.

It is important that this dialogue continues to ensure that members of our Union do everything possible to back up our colleagues responsible for conservation of wildlife and wild places. It is vital that we continue to pursue with vigour this conservation role.'

By the end of the 1980s, IUDZG's commitment to species management through inter-zoo cooperation was firmly rooted. What had started out as a paper on conservation and cooperation at the beginning of the decade had ended up as the heart of the organisation's international conservation strategy. Assisted by IUDZG's efforts and ever-increasing contributions to IUCN/SSC's CBSG, cooperative breeding programmes were now in existence for a considerable number of species and an international programme of zoo collaboration was growing.

Javan gibbon (© Karen Payne)

Participants at the 1993 Annual Conference in Antwerp (© WAZA)

The 1990s

Despite the progress that had been made in favour of conservation in the 1980s, IUDZG's rigid organisational structure held it back from achieving even greater accomplishments. Its restriction of only catering to a small circle of directors from select zoos and aquariums was not suitable for the expanding international scale of the species extinction crisis. In order to help its members accomplish the aims of organising and executing international breeding programmes for endangered species, and to meet the demands of a rapidly changing world, the Union would have to reorganise itself fundamentally.

A far-reaching decision was taken at the 1991 Annual Conference in Singapore to change the Constitution to allow institutions and national and regional zoo associations to become members of the Union, thus significantly expanding the organisation's scope and reach. IUDZG's sweeping organisational change, new name, adopted in 1992 as IUDZG – World Zoo Organisation (WZO), and more focused outlook on what could be achieved through international breeding programmes, poised the organisation to take on even greater conservation challenges from the 1990s onwards.

One of the pinnacle conservation achievements in the history of WZO occurred just two years after the reorganisation. *The World Zoo Conservation Strategy* (WZCS) was published in 1993. Produced together with IUCN/SSC's CBSG, the purpose of the WZCS was to clarify the role of the zoo and aquarium community in global conservation. The writing of the WZCS was largely inspired by the landmark publication *Caring for the Earth: A Strategy for Sustainable Living*, which was published jointly by IUCN, UNEP and WWF in 1991. The WZCS was groundbreaking, signalling the first time that the magnitude of the collective potential of the global zoo and aquarium community had ever been realistically expressed.

Cover of *Caring for the Earth: A Strategy for Sustainable Living*, published in 1991

The Early History of *The World Zoo Conservation Strategy*

By Leobert de Boer, Chair of the European Association of Zoos and Aquaria (EAZA) 2003–2009 and Past Director of the Apenheul Primate Park and GaiaPark

IUDZG decided in 1990 that the zoos and aquariums of the world should have a conservation strategy, just as the botanical gardens had produced (*The Botanic Gardens Conservation Strategy*). Initially the Zoological Society of London offered to take the lead in preparing such a document, but due to the serious financial difficulties that they were facing at the time, they could not afford the time nor staff to do so. The task of drafting the Strategy was accepted in early 1991 by the National Foundation for Research in Zoological Gardens, which later became the European Association of Zoos and Aquaria Executive Office in Amsterdam. This resulted in the completion of a first draft by mid-1991.

Unlike the second edition, which was compiled by twelve or so authors and published in 2005, *The World Zoo Conservation Strategy* (the word 'aquarium' was not mentioned in the title, but the contents clearly included aquariums as specialised zoos) was a single-author document written by Leobert de Boer. Following the production of the first draft, however, several rounds of extensive worldwide consultations were arranged. The successive drafts were sent to all regional associations as well as to a number of national federations asking for comments and input. In addition, WWF and IUCN, including all Specialist Groups, were given an opportunity to comment. Because of this lengthy process the eventual Strategy was not launched until September 1993, unveiled at the WZO/CBSG Annual Conference in Antwerp.

The reactions to the very first draft varied widely, from (literally) 'bullshit' (Ulysses Seal), to 'neither chicken nor meat' (George Rabb) to 'brilliant' (William Conway). The comments on the negative side of the range probably resulted from the fact that some (or many) had believed or wanted that the Strategy would be mainly aimed at convincing the 'outside world', including governments, authorities and fellow conservation organisations, of the important role that zoos and aquariums played in safeguarding endangered species through captive breeding; thus highlighting the successes and stressing the potentials. Instead, the document presented a more comprehensive overview of zoo and aquarium conservation aspects, such as education.

In addition to informing external parties, it also strongly focused on the zoo and aquarium community itself, showing the individual institutions which way to go in order to increase their contributions to conservation.

Halfway through the production of the first edition of the Strategy there was a remarkable incident. Ulysses Seal, Nate Flesness, Tom Foose and George Rabb, the core group of CBSG and ISIS at the time, wanted a more quantitative 'action plan' added. The Strategy's author agreed and a six-page document called *The Ark into the 21st Century* was drafted listing proposed actions and deadlines. The document also included the requirement that zoos should spend 5% of their annual budgets on conservation activities by the end of the 20th century.

The draft action plan was distributed on the occasion of the 1992 World Conference on Breeding Endangered Species in Captivity in Jersey, UK. As a complete surprise to the little group behind the proposed action plan, which included two eventual Heini Hediger Award winners, the WZO Council was furious because the group had dared to define financial requirements for zoos. It was an embarrassing situation for the initiators of the draft action plan. However, how much for the better has the zoo and aquarium community changed since then? Nowadays, less than 20 years later, 5% for conservation is widely considered as hardly enough!

Later in the process, and especially in the years after the publication of the first edition of the Strategy, all agreed that it proved to be one of the most important and influential publications that WZO had ever produced. The document was translated into most, if not all, of the world's major languages and used as a guide to the future by many hundreds of institutions.

One major omission in the original Strategy was already obvious at the time of its publication – it did not enter into the possibilities and necessity of support to *in situ* conservation. This aspect of zoo and aquarium conservation became of ever-increasing importance from the early 1990s onwards and consequently considerable attention was drawn to this in the second edition of the Strategy, which was published 12 years later.

Cover of *The Ark into the 21st Century*, drafted in 1992
Cover of *The World Zoo Conservation Strategy*, published in 1993

Red panda (© Gerald Dick)

Alongside the publication of the WZCS came the understanding that although international breeding programmes were a significant contribution to conservation, they were only a first step. In order for the goals expressed in the WZCS to be achieved, zoos and aquariums would have to develop cooperative collection plans in addition to their cooperative breeding programmes. The thinking of the time was that as animal collections already consumed the largest portion of a zoo's institutional resources, cooperative collection planning and management would deliver significant conservation outcomes without requiring major resource investment. In 1993, WZO President Gunther Nogge, Director of the Cologne Zoo, also noted that:

'Those taxa will survive in our collections for which we will be able to build up self-sustaining populations. Cooperative collection management will increase the diversity of our collections and our potential to fulfil our objectives and to contribute to the conservation of species.'

In order to help members concentrate on matters related to cooperative collection management, WZO took the initiative in 1994 to formally establish a Committee for Inter-Regional Conservation Cooperation (CIRCC). The Committee's main objective would be to facilitate, promote and support the conservation work of the organisation's national and regional member associations from a global perspective. It would represent all of the existing international cooperative captive-breeding programmes. Two years on, WZO members formulated a policy on collection planning that strongly encouraged all member regional associations and institutions to plan their animal collections systematically and collectively.

From the Committee for Inter-Regional Conservation Cooperation (CIRCC) to the WAZA Committee for Population Management (CPM): A Potted History

By Dave Morgan, Chair of CPM and Executive Director of the African Association of Zoos and Aquaria (PAAZAB), Willie Labuschagne, Past Chair of CIRCC and Past Director of the National Zoological Gardens Pretoria, and Jonathan Wilcken, Past Chair of CIRCC and Director of the Auckland Zoo

During the 1990s, leading regional zoo associations established frameworks for coordinating the management of species and animal collections across their member institutions to improve their conservation value. This initiative was overseen by regionally appointed conservation coordinators.

First established by IUCN/SSC's CBSG, the Global Conservation Coordinators' Committee brought together regional conservation coordinators to address the challenge of global planning. Under the oversight of Leobert de Boer, this group was subsequently brought under the auspices of IUDZG and established as a formal committee of that organisation in 1994 as CIRCC. Its role was to coordinate the implementation of global strategies for zoo collection planning and conservation breeding.

In 1996, under the chairmanship of Willie Labuschagne, the Committee focused strongly on promoting cooperation between regions and began a more regular communication between regional associations. Most significantly, the Committee also developed the CIRCC training grants (now adopted as a flagship granting programme of WAZA). This programme was directed towards enabling the transfer of skills and technology between regions to promote effective conservation breeding in zoos worldwide. This granting programme continues today and has distributed US$ 300,000 over the subsequent 15 years.

Bernard Harrison was appointed Chair of the Committee in 1999 and Jonathan Wilcken in 2002. Over this period, CIRCC defined its terms of reference and embarked on a process of establishing WAZA policy positions on records keeping, collection planning and scientific population management. The International Studbook Programme was revised and its administration pulled under the auspices of CIRCC.

In particular, Jonathan Wilcken's administration saw the development of a framework to link regional species management programmes called the Global Species Management Programme (GSMP) and also the establishment of the permanent WAZA staff position of International Studbook Coordinator.

Dave Morgan took on the Chair of CIRCC in 2007, maintaining the population-management genre of the Committee and refining the terms of reference to focus on the International Studbook Programme and what is now the Global Species Management Plan. In keeping with these core activities, the Committee changed its name to the WAZA Committee for Population Management (CPM) in 2008. The next phase of development for CPM will be the full initiation of regional management-programme articulation within the infrastructure of the GSMP.

Further to establishing CIRCC, WZO produced another significant document in 1995, designed to serve as the framework by which the organisation would implement the goals and objectives of the WZCS. The document was called *Zoo Future 2005* and resulted from the first ever WZO Strategic Planning Workshop, which brought together 40 people from all parts of the world, with expertise in the areas of conservation, animal welfare, universities and media. They discussed the future evolution of zoos and the role WZO would play in the process. Participants believed that 'to assume an expanded conservation role, zoos need to become part of a single conservation continuum in which their skills and resources contribute to long-term viability of wildlife in natural habitats'.

Associated with this belief was one of the main components of the work, which was to determine that professional zoos should build cooperative linkages between *in situ* and *ex situ* conservation programmes. While the concept of building cooperative linkages between *ex situ* conservation programmes was not new to the goals of the organisation, the fact that *in situ* conservation programmes should be included was. The critical link to *in situ* conservation programmes had hitherto been provided by IUCN/SSC's CBSG. The fact that WZO was now calling on its members to take on this major task signalled a further evolution in the organisation's commitment to influence conservation.

Cover of *Zoo Future 2005*, published in 1995

1995 Future Search Workshop of WZO

By Gunther Nogge, WZO President 1994–1995 and Past Director of the Cologne Zoo

In 1993, only one year after the first World Summit on the Environment held in Rio de Janeiro, Brazil, WZO together with IUCN/SSC's CBSG had developed its own *World Zoo Conservation Strategy* (WZCS). An Editorial Board consisting of Roger Wheater, Director of the Edinburgh Zoo, Peter Karsten, WZO President, and Ulysses Seal, Chair of CBSG, had led the process in which members from all over the world had contributed their thoughts. In the end the document was brilliantly formulated by Leobert de Boer, Director of EAZA at the time. The WZCS was launched at the WZO Annual Conference hosted by the Antwerp Zoo. At that meeting I was elected President of WZO and my first task was to introduce the Strategy at a press conference at the International Press Centre of the European Community in Brussels, Belgium.

The Strategy received attention both inside and outside of the zoo community. Regional and national zoo associations translated it into various languages. In order to assist the zoo community in implementing the Strategy, the first (part-time) WZO Secretariat was established at the Minnesota Zoo, already home of two other important organisations, CBSG and ISIS. Additionally and with careful preparation, a Strategic Planning Workshop was conducted at the Cologne Zoo in May 1995. The workshop was chaired by two professional facilitators from Canada, Frances Westley and Harrie Vredenburg, who were experienced in management processes related to conservation and environmental protection.

Forty people from all over the world were invited to participate, including not only stakeholders from the zoo community but also outsiders with expertise in areas such as conservation and animal welfare, as well as representatives from universities and the media.

The results of the workshop were laid down in an action plan called *Zoo Future 2005* and was endorsed by WZO members at the 1995 Annual Conference in Dublin. I believe that both the workshop and the document have encouraged zoos all over the world to increase their commitment to conservation. The WZCS and *Zoo Future 2005* are the roots of an ongoing and successful process aimed at turning zoological gardens and aquariums into true conservation centres.

Northern bald ibis (top) (© Johannes Fritz) and Przewalski's horse (© Chris Walzer)

William Conway, Past President and General Director of the Wildlife Conservation Society, added fuel to the fire of investing resources in *in situ* conservation programmes in a keynote speech entitled 'The Changing Role of Zoos in the 21st Century' (Appendix V) given at the 1999 Annual Conference in Pretoria. His speech eloquently called for zoos in the 21st century to become 'proactive wildlife conservation care-givers and intellectual resources'. Further to his speech he added that 'the best of all arks is a functioning ecosystem *in situ* conservation'. William Conway stated that:

'If WZO is going forward, it can't just be seen as getting its own house in order, it must contribute to *in situ* conservation. [WZO's] campaign does more for the zoos and more for the world than just saving wildlife. It places us in a new light as a conservation organisation. It becomes a catalyst, a central organising principle, around which we can all focus our thoughts, as individual institutions, as regional organisations and as a world zoo organisation. I personally think that it will be the one thing that can hold us all together through the next very difficult decade, when we will be under even more pressure from inside and outside our regions.'

Fortified by an inspirational calling, a firm strategic base and an improved organisational structure, WZO was ready to take on the increasing challenges of the 21st century.

The 2000s

Motivated by the keynote address delivered by William Conway in 1999, in the following year members expressed their support for WAZA (renamed in 2000) to consider various ways of participating in *in situ* conservation projects. Through a series of three regional workshops in collaboration with IUCN/SSC's CBSG and hosted by the Cologne Zoo (2000), the Simón Bolivar Zoo (2001) and the Khao Kheow Open Zoo (2001), members examined the feasibility of finding international *in situ* conservation projects that had high public relations, marketing and fundraising values. The aim was for WAZA to align itself with projects that were collaborative and would allow all partners to benefit. Additionally, the projects chosen should make use of the core skills and expertise that were available in zoos, such as small-population management and veterinary skills. What emerged was a list of biological, operational, and institutional and partnership criteria.

Biological Criteria
- Demonstrate a strong link between *ex situ* work in zoos and *in situ* conservation
- Involve one or more globally threatened species
- Involve charismatic species which could function as flagships
- Ensure that any species programme had a positive effect on habitat

Operational Criteria
- Have a strong potential for public awareness and fundraising
- Be based on sound science and management
- Lead to self sustainability in management and funding
- Make use of core skills and expertise available in zoos

Institutional and Partnership Criteria
- Comply with relevant international and national regulations
- Work in partnership with stakeholders
- Gain and have the support of relevant local and national governments and organisations
- Comply with national and international conservation strategies and programmes

WAZA-Branded Conservation Projects

After the WAZA Council had mandated the WAZA Executive Office to initiate the project branding scheme, a pilot phase was started in 2003. A conservation project aiming to protect the Critically Endangered northern bald ibis was the first project to receive the WAZA brand. In the framework of human-led migrations, the project team guides captive-bred animals throughout Europe from the breeding to the wintering grounds. The second project to receive the WAZA brand was associated with protecting the Przewalski's horse and its habitat in Mongolia. Previously listed as Extinct in the Wild, successful reintroductions of captive-bred animals have qualified this species for reassessment to Critically Endangered. Convinced by the successful results of the two pilot projects, the WAZA Council agreed to continue the project branding scheme. Since 2004, the number of WAZA-branded projects has steadily increased, numbering over 200 in 2011.

Development of Branded Projects

The cumulative number of WAZA-branded conservation projects as of December 2011

Year	Projects
2011	216
2010	195
2009	165
2008	147
2007	115
2006	102
2005	71
2004	29
2003	2

Cover of the book *Building a Future for Wildlife: Zoos and Aquariums Committed to Biodiversity Conservation*, published in 2010

A second outcome of the regional workshops was a WAZA *in situ* conservation strategy, which outlined eight key 'Tasks for WAZA Conservation Work':
- To establish a database of *in situ* projects and expertise
- To form an International Conservation Committee
- To develop the WAZA brand for campaigns and projects
- To initiate worldwide campaigns
- To represent WAZA in international legislative forums
- To achieve an international status assisting in seeking funds
- To assist members in communication and collaboration
- To publicise member zoos as centres of conservation

One of the main outcomes of the *in situ* conservation strategy was the initial development of the WAZA brand for conservation projects. It was a system in which suitable *in situ* conservation projects would be awarded the WAZA brand, thereby creating a mutually beneficial situation for both the project and WAZA. The brand itself would promote the project and elevate its status on an international level and the project would allow WAZA to convey what zoos were achieving for conservation globally. It would take another three years before WAZA branding would be implemented.

Parallel to the initial development of the WAZA branding concept in 2000 was the establishment of a new corporate identity for the organisation. As stated by incoming WAZA President Alex Rübel, Director of the Zurich Zoo: 'We, as zoos and zoo associations, will never be respected as global players in conservation if we are not able to speak with one voice and appear under a common brand'. In order to strengthen its influence and credibility as an internationally recognised conservation organisation, a new name was adopted. Later in 2000, instead of being called WZO, the organisation was renamed the World Association of Zoos and Aquariums (WAZA). By including 'aquariums' in its name, the organisation reflected the growing recognition of the importance of aquariums among the membership.

The Value of Aquariums

By Mark Penning, WAZA President 2010–2011
and Director of the uShaka Sea World Durban

Despite 60% of the world's population living within 60 kilometres of the sea and many more living in river catchment areas, only a small proportion of humankind is exposed to aquatic life forms. For most people, their only exposure to aquatic biodiversity is seeing what comes out of a fisherman's net. Aquariums have an enormous role to play in showing us the wonders of our extraordinary and mysterious blue world. Aquariums in one form or another have been popular for centuries. Today there are more than 300 public aquariums worldwide, which host in the region of 200 million visitors each year. The primary role of a public aquarium is to expose the visiting public to the fascinating aquatic environment, and touch the hearts and minds of those visitors and foster in them a true appreciation of aquatic life.

Achieving this depends on creating a balance between exciting entertainment and interaction, between formal and informal education, and the provision of time and space for observation and reflection. But the role of the modern aquarium does not stop there! Aquariums can be powerful drivers of social change and can play a valuable role in conservation by facilitating a change in attitude and behaviour in its visiting public.

Aquariums have seen the need for social change to protect the future of our oceans and lakes and, in turn, our own well-being. Several have developed sustainable seafood initiatives, which help visitors to make sensible seafood choices both in restaurants and in the supermarket. Society is becoming increasingly aware of the burden we place on our planet's resources and by providing accurate and relevant information, aquariums have the potential to influence the shopping habits of millions of people.

Georgia Aquarium (© Gerald Dick)

In the field of science and research, we are in a truly remarkable era. The depths of the oceans have hitherto remained largely unexplored, but tremendous advances in technology now allow us access to some of the fabulous creatures that myths and legends are made of. Dramatic advances in life-support technology, tank design and the use of large acrylic panels instead of glass have allowed aquarium designers far greater scope and more opportunities. Several aquariums can now exhibit the majestic manta ray and the biggest fish on our planet, the incredible whale shark. How many people would ever be fortunate enough to see one of these awesome beasts with their own eyes? There is no denying it – these creatures fascinate us, and our knowledge and understanding have increased immeasurably through the collective efforts of the world's aquariums.

One of the key conservation achievements of 2003 and of the history of the organisation came with the signing of a Memorandum of Understanding (MoU) between WAZA and IUCN. The purpose of the MoU was to enable the Parties to achieve better linkage between the *in situ* and *ex situ* conservation communities. Strengthened by the signing of the MoU with IUCN, and fuelled by the need to encourage and facilitate WAZA members' participation in *in situ* conservation activities, WAZA formed a Conservation Committee in 2003. The Committee's first major task was to oversee the writing of the new WZCS, as the original document was now 10 years old and did not reflect the many developments and new conservation challenges that were facing the zoo community. Jo Gipps, Chair of the WAZA Conservation Committee and Director of the Bristol Zoo, was commissioned to coordinate the writing of the Strategy. The new version was completed in 2005 and was named *Building a Future for Wildlife: The World Zoo and Aquarium Conservation Strategy* (WZACS). Alongside the new Strategy, WAZA published guidelines for its members in order to assist them in becoming involved in *in situ* conservation.

Although increasingly preoccupied by high-level conservation initiatives, such as the signing of an MoU with IUCN and rewriting the Strategy, WAZA had not lost sight of its species-based conservation roots in the new millennium. Alarmed by the extent of the amphibian extinction crisis, for example, WAZA adopted a resolution in 2005 committing itself to 'collaborate with the bodies established by IUCN and its partners to confront the ongoing extinction of amphibians' and encouraging its 'members to join together and contribute'. Kevin Zippel was hired through the joint sponsorship of IUCN/SSC's CBSG and WAZA in 2006 to facilitate the coordination of a global amphibian conservation response. An Amphibian Conservation Planning Workshop was held in Panama the same year to help determine the organisation and policies, species prioritisation, husbandry practices and immediate responses necessary for widespread amphibian conservation. A further resolution was passed in 2007 to support the 'Amphibian Ark' whose mission was to manage amphibian taxa *ex situ* to ensure long-term survival in nature. WAZA committed itself to 'encourage, facilitate, and help fund the practical delivery of *ex situ* programmes'. In 2009 the Amphibian Survival Alliance was established by numerous organisations, joining forces in order to implement the IUCN *Amphibian Conservation Action Plan*. Focus of the newly established initiative was habitat conservation, combating disease and dealing with over-harvesting. The effective start of work was on 1 June 2011 when an Executive Director and a Chief Scientist took up their positions. The donors of the initial phase were the Zoological Society of London, George Rabb, Chester Zoo, Conservation International, Detroit Zoological Society, Wildlife Conservation Society and Frankfurt Zoo.

Another species-focused resolution was passed in 2007 regarding the conservation of India's Critically Endangered gharials. WAZA dedicated itself to 'liaise and collaborate with the relevant bodies established by the IUCN and its partners to confront the ongoing extinction' of the species.

Indian gharial (© Gerald Dick)

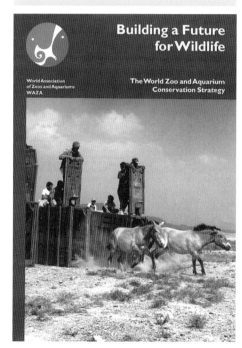

Cover of *Building a Future for Wildlife: The World Zoo and Aquarium Conservation Strategy*, published in 2005

WAZA's Involvement in the Amphibian Ark Initiative

By Chris West, Chair of the Amphibian Ark and CEO of Zoos South Australia, and Lena Lindén, Chair of the Amphibian Ark and CEO of the Nordens Ark

The first intimation of a creeping catastrophe of amphibian extinction can be traced back to the 1990s and earlier. Practitioners of field research had pieced together a bleak image of disappearance trends and, by the early years of this century, it was increasingly clear that a combination of threats, notably habitat shrinkage and fungal disease (chytridiomycosis), were so dramatic that urgent large-scale action was imperative. Moreover, that such concerted action would be most effective if the perceived separation of *ex situ* and *in situ* conservationists was set aside with a recognition of interdependence. With 1,300 zoos worldwide spending about US$ 350 million a year on wildlife conservation and visited by over 700 million people, holding at least 40,000 amphibian specimens, the zoo community has enormous potential and capacity to make a difference.

The Amphibian Ark (AArk) came about through discussions at multi-disciplinary and organisational workshops because of a manifest existential threat to a whole vertebrate taxon. It was founded in 2006 and has three parent organisations, WAZA, IUCN/SSC's CBSG and the IUCN/SSC Amphibian Specialist Group. It has a mission to ensure the global survival of amphibians with the emphasis on those that are likely to disappear in the wild. The concept of AArk is that it is given its strength through the collective action of a global community that already exists, namely zoos. Its activities encompass population support through secure breeding, relevant research, fundraising, raising public awareness, and lobbying for shifts in policy and political priority. AArk is a classical umbrella organisation with a very small and cost-effective administrative hub.

Key activities of AArk over its relatively short lifespan include the leadership of the Year of the Frog in 2008, increasing *ex situ* capacity and capability, and providing support to partners that are working in the field in support of wider *Amphibian Conservation Action Plan* activities.

The Year of the Frog was conceived as an exercise to raise awareness and money using campaign methods pioneered by EAZA. It was the first global campaign run by the zoo community and it attracted enormous attention and very high-level support from people such as David Attenborough, Jeffrey Corwin, Jean-Michel Cousteau and Jane Goodall. Kermit the Frog was loaned from Disney to spread the message and in Europe two real live princesses were enlisted to kiss frogs and take part in media events. Corporate sponsor Clorox provided financial support in the field in El Valle de Antón in Panama, and many zoos and individuals around the world achieved remarkable results. In retrospect, the successes were perhaps more in terms of public awareness than actual raising of funds for specific programmes, which is a reflection of the challenging environment for conservation campaigns. It demonstrated what zoos can do if they mobilise themselves. So many WAZA members went beyond interpretation and public education, and established facilities and resources to support specific species and locations.

Princess Xenia of Saxony kissing a frog at the Leipzig Zoo during a Year of the Frog event in 2008 (© Leipzig Zoo)

To ensure ongoing public awareness, the AArk team has created a free membership programme. Members receive a quarterly newsletter to update them on major developments in amphibian conservation. Over 5,500 members have subscribed to the AArk newsletter at the time of writing and this number continues to grow.

Also ongoing is the building of greater capacity to expand *ex situ* activities to ensure a wider, firmer base for rescue of amphibians from extinction. Priority actions have been assessed and agreed upon, partners have been identified and secured, skills have been transferred and guidelines have been put in place.

A process of AArk Conservation Needs Assessment Workshops has been managed on a country basis considering species needs and with specific actions as an output. From 2006 over 37% of the world's amphibians have had their needs assessed in this fashion. Once the needs are affirmed a further exercise of conducting AArk Training Workshops follows and, since 2004, over 1,400 people have learned about amphibian biology and conservation techniques in a total of 46 courses, held in 28 countries.

A number of key areas of management have been defined in guideline documents, such as population management, reintroduction, husbandry in modified containers, disease management, bio-banking and research. AArk facilitates a matching of those amphibians most in need of support, and the zoos and other organisations which are most able to provide effective support through international-level partnerships. Many zoos have stepped up to the plate but many projects remain unadopted.

One way in which WAZA members have been supportive of AArk is via grant support. Currently over 24 schemes are operating, which extend beyond *ex situ* activities. This enables the leveraging of further funds and in-kind support. Many innovative ideas have also been explored, such as auctioning the naming rights of newly found amphibian species.

In summary, AArk, with its very significant support from WAZA, has made a vital contribution to the conservation efforts to save amphibian species that are facing extinction in the wild. It has fostered a new set of partnerships and raised very significant public awareness in the plight of a whole vertebrate taxon. It has also raised money and established professional standards and practices spanning zoos to field operations and research laboratories. It serves as a model for the all too many taxa following amphibians down the track to extinction and it has rested on the translation of policy and inter-organisational collaboration into action via a small group of dedicated expert conservationists.

Particular credit must be given to Kevin Zippel for providing the energy and leadership that have inspired many zoo employees to have a passion for amphibians and conservation. Credit must also be given to the founding impetus from a CBSG meeting and to colleagues from WAZA who have acted in an AArk support capacity, namely Jeffrey Bonner and Gordon McGregor Reid.

It has to be said that the work of AArk is unfinished. It has only really just started and many, if not most, amphibians still face extinction. Many WAZA members have committed to the cause of conservation as it is central to the role and purpose of modern zoos but many more can still take part. In fact they must.

Mazán harlequin frog (© Luis Coloma)

In 2008, WAZA and IUCN extended their MoU for a further five years. Dedicated to collaborating with more international partners to achieve its conservation objectives, WAZA broadened its reach in the conservation community by signing a series of MoUs with other leading conservation partners between 2008 and 2009. WAZA and the Secretariat of the Convention on Migratory Species (CMS) signed an MoU in 2008, realising that activities undertaken by WAZA concern migratory species and recognising that they 'pursue common goals in the conservation of ecosystems and the protection of migrating species, which can only be successfully met by enhanced and concerted actions on different levels and between all sectors'.

In 2009, WAZA signed an MoU with the Secretariat of the Convention on Biological Diversity (CBD), aiming to use the MoU 'as a general framework and guiding tool in identifying and carrying out specific collaborative projects and activities between the Parties'. The same year, WAZA signed an MoU with the Secretariat of the Ramsar Convention on Wetlands, 'recognising that zoos and aquariums are occupied with wetland-related species as part of their exhibits, education work as well as field conservation projects'.

During the 2009 Annual Conference in St. Louis, a Climate Change Task Force was established, jointly with IUCN/SSC's CBSG. A total of 788 signatures of senior management of the world's leading zoos and aquariums were collected in support of a petition sent to the United Nations Secretary General to reduce atmospheric carbon dioxide emissions to the safe level of less than 350 parts per million.

Extension of MoU between WAZA and IUCN in 2008, left to right: Julia Marton-Lefèvre, IUCN Director General, Jane Smart and Dena Cator of IUCN Species Programme and Gerald Dick, WAZA Executive Director (© Lynne Labanne/IUCN)

WAZA's collaboration with international partners continued in 2009 with the implementation of the Year of the Gorilla. The concerted international campaign highlighted the conservation status of gorillas and was developed in cooperation with CMS, the United Nations Environment Programme (UNEP), the United Nations Educational, Scientific and Cultural Organisation (UNESCO), the Great Apes Survival Partnership (GRASP) and the United Nations Decade of Education for Sustainable Development. Additionally, over 100 WAZA member zoos participated in the event through a wide variety of zoo-based activities. In 2011, WAZA supported the CMS campaign 'Year of the Bat'; numerous members organised educational events and supported conservation projects. Another significant conservation achievement of 2009 was the publication and launch of the first global aquarium strategy for conservation and sustainability called *Turning the Tide*. Developed by and for aquarium professionals and partners, the strategy described the essential role that aquariums can play in furthering the conservation of aquatic species.

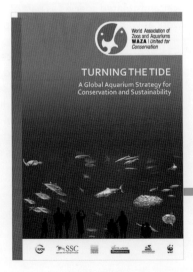

Cover of *Turning the Tide: A Global Aquarium Strategy for Conservation and Sustainability*, published in 2009

Turning the Tide

By Mark Penning, WAZA President 2010–2011 and Director of the uShaka Sea World Durban

Whilst many internationally respected individuals from public aquariums contributed to the WZACS, it was recognised that there are important differences between aquariums and zoos in their nature, character, typical constitution, operational requirements, in their stakeholder community and in the challenges faced. Hence, it was decided that a dedicated publication was required to detail carefully the implementation of the WZACS by public aquariums.

Launched during the 2009 Annual Conference in St. Louis, *Turning the Tide* drew on the expertise of WAZA institutional and regional association members, together with colleagues from the International Aquarium Forum (IAF), International Aquarium Congress (IAC) and European Union of Aquarium Curators (EUAC). Through this cooperative, strategic approach the aquariums of the world unite to make a meaningful contribution to global efforts in aquatic conservation and sustainability. The document is formally endorsed by the IUCN/SSC, Ramsar Convention on Wetlands, Conservation International, Wetlands International and World Wildlife Fund-US.

Turning the Tide plots the course for public aquariums in a world where marine, coastal and freshwater resources are being ruthlessly exploited, where water-associated biodiversity is steadily declining and where the careful management of all aquatic ecosystems is crucial for the well-being of the planet. With unprecedented human impacts evident across the globe, scientists point out that increasing ocean surface temperatures and ocean acidification may make the survival of coral reefs the greatest conservation challenge of our times. The implications are substantial in terms of species extinctions, loss of livelihoods for those human communities dependent on reef-associated fisheries and decreased coastline protection with the risk of catastrophic floods.

For this strategy to be meaningful, the aquarium community needed to translate it into languages relevant in areas of the world most under threat; Spanish and simple Chinese versions were launched during 2010, and the Japanese version was launched at the Convention on Biological Diversity's 10th meeting of the Conference of the Parties (CBD CoP10) in Nagoya, Japan, in 2010. Work continues on further translations into languages of South East Asia.

As a community, we now know what messages we need to give our visiting public and how to empower them to make the changes needed to ensure our survival on a healthy planet. The next phase will be to assess how effective we are at facilitating those changes.

2010 International Year of Biodiversity

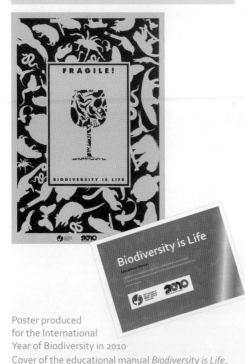

Poster produced for the International Year of Biodiversity in 2010

Cover of the educational manual *Biodiversity is Life*, produced jointly by WAZA and IZE in 2010

United Nations Decade on Biodiversity

The year 2010 was a similarly eventful year for WAZA. The organisation was selected by CBD as 'Partner of the International Year of Biodiversity'. WAZA member zoos undertook a wide range of activities on their grounds and the WAZA Executive Office contributed a variety of products, such as posters, the education manual *Biodiversity is Life*, produced jointly by WAZA and IZE, and a book entitled *Building a Future for Wildlife: Zoos and Aquariums Committed to Biodiversity Conservation*. The book's main focus is on 25 conservation success stories from around the world, all drawn from the WAZA-branded projects, detailing the conservation achievements of zoos and aquariums. The products were designed to:

- Enhance public awareness on the importance of conserving biodiversity and on the underlying threats to biodiversity
- Raise awareness about the accomplishments to save biodiversity by communities and governments and to promote innovative solutions to reduce threats to biodiversity
- Call on individuals, organisations and governments to take immediate steps to halt biodiversity loss
- Initiate dialogue among stakeholders on the necessary steps for the post-2010 period

Another major achievement for WAZA was to receive external grants for its conservation projects. The first grant was for US$ 25,000 from the Mohammed bin Zayed Species Conservation Fund to support five WAZA-branded projects. The second grant of CHF 50,000 came from the Swiss-based MAVA Foundation to support 'WAZA's worldwide initiative to drive biodiversity – 2010 and beyond'. In 2011, WAZA signed MoUs with CBD for the United Nations Decade on Biodiversity (2011–2020), the Alliance of Marine Mammal Parks and Aquariums (AMMPA) and CITES.

WAZA's Relationship with IUCN

The International Union for Conservation of Nature (IUCN) is an international organisation whose members are sovereign states, government departments, non-political bodies and international entities. Its broad purpose is to foster the maintenance of the biosphere and its diversity by rational management of the Earth's resources. It urges and assists the development and application of wiser policies on conservation, both at the technical and broader socio-political levels. In the course of its programmes, it seeks to halt all forms of deterioration as well as to promote the conservation of wild places, and wild animals and plants in their natural habitats. IUCN was founded in 1948 as the International Union for the Protection of Nature (IUPN) at an international conference at Fontainebleau, France, sponsored by UNESCO and the government of France. IUDZG was one of its founding members.

Excerpt of the IUCN founding document, with Achille Urbain as IUDZG representative (© IUCN)

IUCN logos over time

Participants at the 1974 Annual Conference in Basel (© WAZA)

IUDZG became a member of IUCN in 1949. There were not enough funds to pay the US$ 50 membership fees the first year and an unnamed member is famously quoted as stating that: 'As the Union has no financial means, the support of the International Union for the Protection of Nature will be purely moral'. A member zoo, which remains anonymous, agreed to pay the following year's dues on behalf of IUDZG. The relationship between IUDZG and IUCN took time to develop. Progress was slow and in the early days IUDZG struggled to gain acceptance into the conservation community, which, according to IUDZG Secretary Charles Schroeder, Director of the San Diego Zoo, in a 1963 letter to IUDZG President, George Mottershead, Director of the Chester Zoo, was due to a 'lack of interest in zoological gardens, or lack of knowledge of zoos and their role generally'.

Understanding that there was a need to form a link between itself and the greater zoological community in order to further its conservation objectives, IUCN called for the formation of a 'Zoo Liaison Committee' in 1963. Besides forming a link, a principal function of the Committee would be to collect, collate and distribute information on breeding rare animals in captivity, and to arrange a regular exchange of information between zoos and wildlife departments. Responding to IUCN's request for cooperation, IUDZG passed a resolution in 1964 that 'an international organisation of zoological gardens, animal collections, experimental research institutions and animal traders should be developed through the agency of the IUCN Zoo Liaison Committee; that this federation should be formed, among other purposes, for developing a method of distributing certain rare animals for exhibition in cooperation with IUCN Survival Service Commission (SSC) and for cooperating with the IUCN and various national governments in the administration of this method'.

Participants at the 1967 Annual Conference in Barcelona (© WAZA)

Although progressive in its mission and valuable as a tool for inter-agency cooperation, the existence of the Zoo Liaison Committee was short-lived. Shaky political underpinnings, a general lack of effectiveness and no prospect of funds to pay for any support staff caused the Committee to fold in 1973. Understanding that a vital link between zoos and IUCN was still necessary, however, IUDZG members resolved that they would continue to strengthen ties with SSC in order to win a seat on the Commission.

Participants at the 1977 Annual Conference in Vienna (© WAZA)

One of the main objectives for IUDZG in the early 1970s was to become formally recognised by IUCN as the official international zoo organisation. According to IUDZG's Second Vice-President Colin Rawlins, Director of Zoos for the Zoological Society of London in 1974, IUDZG did not want to be 'regarded as only one body among many others whose views on zoo matters should be listened to'. In 1976, IUDZG member Ernst Lang, Director of the Basel Zoo, had a meeting with IUCN's Executive Director and the Administrative Officer to pursue IUDZG's claim for recognition. He asked why IUDZG was not 'officially recognised by IUCN as *the* international body representing good zoos'. The response was quite simply, 'what is a good zoo?'. IUCN's blunt response prompted IUDZG to discuss the need to 'fix criteria which are used to define a good zoo', thereby generating the first debate on the question of individual versus institutional membership. The prevailing thinking was that as long as some of the world's major zoos were excluded from membership, it would be difficult for IUDZG to be accepted universally as the only international zoo body. It would take another 15 years before IUDZG would change its membership structure to become truly representative as an international zoo organisation.

By 1977, IUDZG's goal to seek exclusive recognition by IUCN was dampened. After some discussion at that year's Annual Conference in Vienna, it was agreed that IUDZG should not seek exclusive recognition by IUCN but should 'continue to do the best possible conservation job and be represented in IUCN'. IUDZG members decided that the fullest cooperation should be offered to IUCN and a direct liaison should be maintained. Furthermore, members confirmed that IUDZG 'would not seek to be regarded as a body representing all zoos in the world, but as the only international zoo body in existence and representative of the major non-commercial zoos of the world'.

By the 1980s, the critical state of world conservation and the rapid increase in the list of threatened species prompted the relationship between IUCN and IUDZG to progress. Having recognised that 'zoological and botanical gardens can have an important role to fulfil in the conservation of wild species' in a 1975 resolution, IUCN submitted a policy statement in 1980, outlining principles and recommendations for the keeping of wild animals in captivity. It was the first policy statement written by IUCN directly related to zoos. It recommended that captive-breeding programmes be coordinated in the interests of the survival of species.

Alongside the submission of the policy statement, IUCN renamed SSC the 'Species Survival Commission' (instead of the Survival Service Commission) and formed the Captive Breeding Specialist Group (CBSG, now the Conservation Breeding Specialist Group). CBSG was created by IUCN/SSC to provide a liaison among Specialist Groups and members of the captive-breeding community, and to provide advice to SSC on captive-breeding matters. It emerged from the preceding Zoo Liaison Committee and its formation provided a direct link to the captive-breeding work that IUDZG member zoos were undertaking, bridging the gap between the two organisations.

Director Walter Fiedler leading a zoo tour at the 1977 Annual Conference in Vienna (© Vienna Zoo)

WAZA and the Conservation Breeding Specialist Group (CBSG): An Increasingly Productive Conservation Collaboration

By Onnie Byers, CBSG Executive Director 2005–2011, and Robert Lacy, CBSG Chair 2003–2011

The CBSG is an international conservation network dedicated to saving threatened species. We use innovative, scientifically sound processes to bring people together to effect positive conservation change. CBSG is a part of the Species Survival Commission (SSC) of IUCN and this relationship is essential to the strength of our organisation.

Originally the Zoo Liaison Committee when formed in the early 1970s, CBSG's mission was to provide an interface between IUCN and zoos, and CBSG was instrumental in developing and promoting the scientific management of captive wildlife populations. CBSG in its present form began with the appointment of Ulysses Seal as Chair in 1979 and continues under the current Chair (since 2003), Robert Lacy.

As species in the wild increasingly began to require the same kinds of intensive management as animals in captivity, CBSG expanded its scope to small-population management and linking *in situ* and *ex situ* scientific expertise. We moved from having a focus on captive populations to an organisation with a much broader approach to tackling conservation problems in partnership with zoos and non-zoo organisations.

One of CBSG's strengths has always been bringing a scientific approach to defining problems and determining management strategies for conservation in captivity and in the wild. A major turning point came in 1985, when CBSG became involved with the recovery programme for the black-footed ferret.

This effort pioneered the application of population-biology methods, including computer modelling and the use of expertise from the zoo community, to help guide recovery planning for species on the verge of extinction in the wild. This approach was further developed by CBSG and became the Population and Habitat Viability Assessment (PHVA) workshop methodology.

Ulysses Seal, CBSG Chair 1979–2003 (© CBSG)

CONSERVATION BREEDING SPECIALIST GROUP

The development of the VORTEX programme, by Robert Lacy of the Chicago Zoological Society, was pivotal in moving CBSG programmes forward. This small-population biology tool was integrated into our workshops, further expanding the PHVA process. Combining field and captive data and expertise, PHVAs continue to provide a unique forum in which wildlife managers, academics and captive-breeding experts can work together on risk-assessment, species management and recovery planning. Under Robert Lacy's leadership, species risk-assessment tools have been improved and expanded into meta-models allowing VORTEX to interact with models of other threatening processes.

Under its non-profit financial arm, the Global Conservation Network, CBSG receives annual contributions from zoos, aquariums, zoo associations and individuals around the world. WAZA is one of CBSG's major sponsors. However, CBSG does not have a formal connection to WAZA or to any zoo association. As an SSC Specialist Group, CBSG plays a unique role within IUCN as a neutral link between the *ex situ* community and species conservation efforts. Maintaining independence and neutrality is important so that CBSG can serve as a facilitator of discussions and collaborations between organisations with sometimes divergent perspectives and goals.

While closely aligned, WAZA's role as a professional organisation representing its members can seem at odds with CBSG's role as an independent network of individuals serving the SSC. However, throughout the years, CBSG and WAZA have worked together on important conservation initiatives to which the complementary strengths of each have contributed significantly. Our relationship goes back a long way and today is stronger and more productive than ever.

WAZA has long been the coordinating organisation for international studbooks, the key public record of the history of zoo breeding programmes. These and regional studbooks provide the basis for cooperative, scientific management of breeding programmes. In the late 1970s and 1980s, CBSG, together with colleagues in ISIS, zoo associations, individual zoos and academia, developed techniques and tools for using studbooks to guide breeding and transfer decisions.

As major zoos were struggling to find ways to strengthen wild-animal breeding in the early 1980s, William Conway and Theodore Reed led an effort to create a 'Wild Animal Propagation Trust'. The effort was followed by many other meetings, including one on the 'Scientific Justification of Studbooks' in Copenhagen, Denmark (which included Katherine Ralls and Devra Kleiman). Eventually, William Conway and Tom Foose wrote the first 'Species Survival Plan' (SSP). After widely circulated drafts, the Plan was published in 1983.

Ulysses Seal, working with Tom Foose, from what is now AZA, made a major contribution by writing the Siberian tiger SSP species programme, the first approved by AZA. The resulting analyses and breeding recommendations were used as a model for many subsequent SSPs. Following the proliferation of SSP-type programmes in other regions, the Global Conservation Coordinators' Committee was created to encourage the scientific planning and management of zoological collections, regional training and the contribution of members to in situ projects. This Committee evolved into the Committee for Population Management (CPM) with oversight of international studbooks and Global Species Management Plans (GSMPs), under the auspices of WAZA.

The publication of the groundbreaking World Zoo Conservation Strategy in 1993 was the result of a close collaboration between CBSG and WAZA. This document articulated a vision of the role of zoos and aquariums in conservation for the next 10 years. WAZA produced a new strategy in 2005, with CBSG again playing a significant role in its conceptualisation and development.

Another excellent collaboration between CBSG and WAZA was the formation in 2006 of the Amphibian Ark (AArk) in response to the amphibian extinction crisis. AArk is a collaboration among zoological institutions, the IUCN/SSC Amphibian Specialist Group, other conservation organisations and private breeders who collectively strive to protect those species of amphibians that cannot currently be safeguarded in nature. WAZA serves as the essential link and advocate to its members, and CBSG provides the connection to individual expertise and broad collaborative networks.

Recently, CBSG, WAZA and other partners in the zoo community have begun work to identify how intensively managed populations (both ex situ and in situ) can be managed more effectively to serve species conservation.

Siberian tiger (© Christian Schmidt)

CBSG has evolved significantly over the years. From an office of one, we have grown to an office of six (headquartered at the Minnesota Zoo) with a volunteer network of over 500 professionals and nine regional networks. CBSG champions openness, inclusiveness, ethics and risk-taking, and will continue to evolve in response to the needs of those concerned with conserving the planet's biodiversity.

WAZA has evolved just as significantly since its inception. The changes occurring in each organisation provide ever-increasing opportunities for productive partnerships to save species, and the bonds between WAZA and CBSG have strengthened with each collaboration. CBSG believes that this strengthening cooperation will increase the integration of in situ and ex situ species management and ultimately ensure long-term species survival. We look forward to new and continuing collaborations with WAZA in the years ahead.

In 1981, IUDZG donated US$ 1,000 to IUCN/SSC's CBSG in the hope that closer links could be developed between the two organisations and IUDZG could increase its influence on international decisions about wildlife. Funds were given to support the operation of the CBSG office, the cost of travel and secretarial support. Initially, IUDZG was cautious about donating future funds to CBSG, wanting first to be convinced of the value of the group's contributions and of the effectiveness of their approach. Five years on, thanks largely to the work of CBSG, coordination between zoos had improved and a variety of zoo breeding programmes had been developed. IUDZG made a second, more substantial donation to CBSG in 1986 and contributions have continued ever since.

In 1987, IUCN approved a policy statement on captive breeding, urging that 'those national and international organisations and those individual institutions concerned with maintaining wild animals in captivity commit themselves to a general policy of developing demographically self-sustaining captive populations of endangered species wherever necessary'. The essence of the policy statement later became part of IUCN's 1991 conservation strategy *Caring for the Earth: A Strategy for Sustainable Living*, which was a successor to the original *World Conservation Strategy* published in 1980. The new publication was developed to be used by 'those who shape policy and make decisions that affect the course of development and the condition of our environment'. It emphasised that the highest priority for the conservation of biological diversity was *in situ* conservation of species in their natural habitats. Among its suggestions for species conservation, it stated that 'zoological gardens have a key role in maintaining *ex situ* populations of animals' and highlighted that all zoos should join the network established by IUCN/SSC's CBSG.

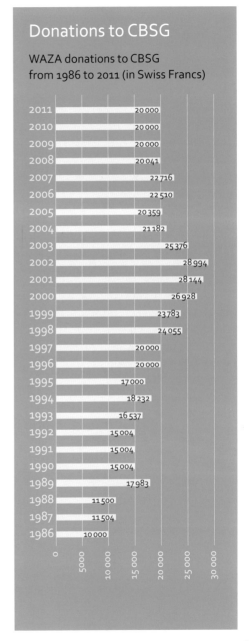

Donations to CBSG

WAZA donations to CBSG from 1986 to 2011 (in Swiss Francs)

Year	Amount
2011	20 000
2010	20 000
2009	20 000
2008	20 041
2007	22 716
2006	22 510
2005	20 359
2004	21 182
2003	25 376
2002	28 994
2001	28 144
2000	26 928
1999	23 783
1998	24 055
1997	20 000
1996	20 000
1995	17 000
1994	18 232
1993	16 537
1992	15 004
1991	15 004
1990	15 004
1989	17 983
1988	11 500
1987	11 504
1986	10 000

WAZA's New Home

As of 1 May 2010, the WAZA Executive Office became situated in IUCN's newly built Conservation Centre in Gland, Switzerland. For WAZA it offers the opportunity to better liaise with the conservation community and use the synergies for cooperation and office management. IUCN holds that:

'The Conservation Centre will serve not only as a place of work for [IUCN's] global Secretariat team, but will also serve as a hub for global collaboration with the aim of conserving nature. The new Conservation Centre will enhance [IUCN's] leadership role as an international forum to cultivate alliances and partnerships for stronger collective action among the conservation community, government and society.'

IUCN Conservation Centre in Gland (© Holcim)

By 1988, the work accomplished by CBSG had, according to IUDZG member Jeremy Mallinson, Director of the Jersey Wildlife Preservation Trust, 'raised the image of zoos and [had] gained support and considerations for the achievements made in conservation'. Long-standing CBSG Chair Ulysses Seal further captured the general feeling of the time in a 1989 report:

'[I cannot] overemphasise the turnaround that [I] have experienced in the conservation community towards captive propagation. It has turned from active hostility, distrust and distaste to not only active recognition, but to requests for the participation of the captive propagation community to preserve a wide range of species.'

A close and mutually beneficial relationship between CBSG (IUCN by extension) and WZO continued throughout the 1990s. The relationship became particularly fruitful when WZO took the pivotal decision in 1991 to broaden its membership by including institutions, and national and regional associations, thus providing a link between more individual zoos and CBSG, and widening its sphere of influence. Working jointly, WZO and CBSG launched *The World Zoo Conservation Strategy* (WZCS) in 1993. The Strategy represented a major milestone both for the zoo community and for the solidification of the relationship between WZO and CBSG. As progressively more species in the wild came to require the same kinds of intensive care that animals require in captivity, CBSG expanded its range in the mid-1990s to include small-population management and the linking of *in situ* and *ex situ* scientific expertise. WZO benefited from CBSG's proficiency in this domain as it expanded into *in situ* based conservation at the start of the new millennium.

The biggest milestone in the history of WAZA's relationship with IUCN came in 2003 with the signing of a Memorandum of Understanding (MoU) between WAZA and IUCN. The signing of the MoU was the result of over 20 years of productive collaboration with CBSG and its purpose was to enable the Parties to achieve better linkages between the *in situ* and *ex situ* conservation communities.

The signing of the MoU intensified cooperation between WAZA and CBSG considerably. Regular correspondence and exchange of publications between the two organisations took place, and WAZA's financial contributions to IUCN increased. In addition to its yearly payments to CBSG, WAZA now helped to sponsor (and still continues to do so) some of the running costs of the office of the Chair of IUCN's SSC and it contributed funds to specific Specialist Groups. As WAZA's interests inside IUCN were becoming more diversified, WAZA requested that its focal point be changed from CBSG to the IUCN headquarters. In 2007, under the leadership of IUCN's Director General Julia Marton-Lefèvre, WAZA's new primary contact became IUCN's Species Programme. A year later, WAZA's MoU with IUCN was renewed for a period of five more years and, momentously, WAZA was invited to move its Executive Office into the IUCN World Headquarters in Gland, Switzerland.

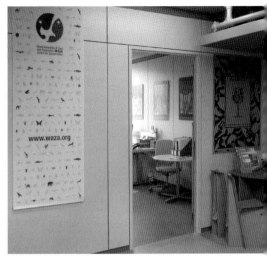

Since 2010, the WAZA Executive Office is located in the IUCN Conservation Centre in Gland (© WAZA)

The International Studbook Programme

European bison (© Christian Schmidt)

When after World War I, it appeared that the European bison was in danger of extinction, the holders of the species met at the Berlin Zoo in 1923 to form the 'International Society for the Protection of the European Bison'. The European Bison Studbook, the first for a wild species and modelled on those in use for domestic animals, was set up and came into effect in 1932.

A studbook is a register that lists and records all captive individuals of a species. The main purpose is to facilitate the continued and coordinated breeding management of the species. Without such a record, it is often impossible to trace the past history of an individual. Properly run and containing the right data, studbooks can be used to provide the genetic and demographic guidelines necessary for managing reproduction for both the increase and the limitation of the population in question. Studbooks can also be used to help recognise detrimental inbreeding effects.

By 1956, despite being aware of the growing need to coordinate information about rare animals in captivity, IUDZG was not in a financial position to advance the effort, as affirmed by IUDZG President Walter Van den bergh, Director of the Antwerp Zoo: 'It is not enough to collect facts, dates, etc. in connection with the proposed aim: IUDZG has to publish this information, this means an expenditure of money. This will not be possible as long as the organisation is not better organised and in a more favourable financial position'.

It was not until 1964, at a symposium on 'Zoos and Conservation', held in London, UK, organised by IUCN, IUDZG and the International Council for Bird Preservation (now BirdLife International) that it was officially recognised that studbooks should be started for certain rare animals:

'Where certain species are concerned, studbooks will be essential. If a species is to be bred in captivity over a number of years, it is important to know the degree of inbreeding occurring; while in the absence of natural selection, artificial selection of healthy breeding lines is vital.'

Two years after the symposium, IUCN/SSC and IUDZG officially established studbooks, then known as 'The Stud Book of Rare and Threatened with Extinction Animals, Bred in Zoological Gardens'. At a meeting of IUCN/SSC in 1966, a number of recommendations were made for the organisation of studbooks, principally that the central organising body should be the IUCN/SSC Zoo Liaison Committee. The recommendations were welcomed by IUDZG.

The first rules and procedures for the establishment and maintenance of studbooks, and the responsibilities of studbook keepers, were drawn up by a working committee of IUCN/SSC's Zoo Liaison Committee. In 1970, the Council of the Zoological Society of London agreed to a proposal from IUCN/SSC that the Society be responsible for the coordination of international studbooks and that the editor of the *International Zoo Yearbook* should undertake this task. By the early 1970s, the work of the Zoo Liaison Committee, including its studbook overseeing role, had been superseded by IUCN/SSC's recently formed CBSG. In 1975, Peter Olney, Curator of Birds at the Zoological Society of London, took over as the editor of the *International Zoo Yearbook* from Nicole Duplaix-Hall, and in that capacity inherited the role of International Studbook Coordinator.

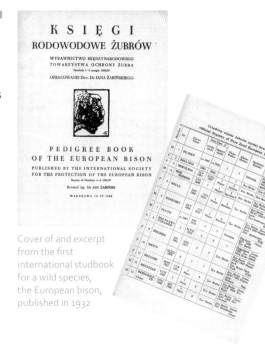

Cover of and excerpt from the first international studbook for a wild species, the European bison, published in 1932

First list of international studbooks published in volume 6 of the *International Zoo Yearbook* in 1966

International Studbooks

By Peter Olney, International Studbook Coordinator 1975–2003

In 1975, I was asked to take on the editorship of *International Zoo Yearbook* and one of the tasks of the editor was the coordination of international studbooks; this included presenting an annual report to IUDZG and IUCN/SSC. These reports summarised the status of each studbook, proposed new studbooks, and raised any other matters relating to the collection and compilation of studbook data. At that time there was considerable debate on the ethics of keeping wild animals in zoos and the subject was often discussed within both international bodies and their committees. These discussions were occasionally vociferous and emotional, especially at SSC meetings, but under the benign chairmanship of Peter Scott (1963–1980), everyone was always dealt with fairly and tactfully, and I do not remember any proposal for a new international studbook being rejected.

I had known and admired Peter Scott for many years since the time I joined the staff as a research biologist; he became my employer, mentor and friend. It was therefore a bonus for me in those early years to have Peter Scott as Chair of SSC. I was equally fortunate in having the charismatic and innovative Ulysses Seal as Chair when from 1979 I reported directly to CBSG.

The potential value of studbooks should be appreciated even by those opposed to zoos. Studbooks remain the essential tool in any coordinated and scientifically managed captive population; the data can be used to construct pedigrees and lineages, and the genetic and demographic assessment, and manipulation of data are fundamental to the success of such programmes.

One of the best examples of a long-running studbook that has provided the basic data for use in a successful international conservation programme, and which includes a reintroduction programme based on the captive population, is that of the endangered golden lion tamarin of south-eastern Brazil. An example of a relatively new international studbook that has the potential to save a species now presumed extinct in the wild is that of the Spix's macaw, also from Brazil. The data collected on the small number of living individuals (68 registered in 2009) and analysed by the studbook keeper is impressive in its detail and the recommendations for the management of the captive-breeding programme from the studbook keeper and coordinator give some hope for the survival of this beautiful bird.

In 1979, IUDZG sponsored a symposium on 'The Use and Practice of Wild Animal Studbooks' in Copenhagen, Denmark. At the time, IUDZG was eager to assume the leading role in the world for the management of wild-animal studbooks. The sentiment was exemplified by IUDZG President Colin Rawlins, Director of the Zoological Society of London, who saw the symposium as an 'opportunity to confirm IUDZG's position as an authoritative international body in the field of wild animal captive breeding'. In order to achieve its aim, IUDZG spent a record-breaking CHF 10,000 on the event, representing the largest expenditure ever made by the organisation at the time.

The symposium led to the first major amendments of the rules for the establishment and maintenance of studbooks. A list of 15 IUDZG resolutions also resulted from the event. Three of the most noteworthy included the following:
- IUDZG should prepare a priority list of critical species to be maintained in captivity
- IUDZG should prepare long-term international breeding programmes for the listed critical species and encourage all zoos to follow scientific principles and ideals in the management of their stocks of their species
- The appointment of studbook keepers should remain the responsibility of the International Studbook Coordinator with the advice and endorsement of IUDZG and IUCN/SSC

Golden lion tamarin (© Christian Schmidt)

Throughout the 1980s, the number of studbooks continued to grow. In 1994, WZO's newly formed Committee for Inter-Regional Conservation Cooperation (CIRCC) accepted overall responsibility for international studbooks. The rules and procedures have been reviewed and updated on a number of occasions, particularly in 2003, with major reviews in 2007 and 2010. Since then the WAZA Executive Office has taken on the overall responsibility and coordination of all international studbooks, appointing its own International Studbook Coordinator. As of December 2011, there were 124 active international studbooks, which include 163 species or subspecies.

Spix's macaw (© Christian Schmidt)

History of the
International Species Information System (ISIS)

By Nate Flesness, ISIS Science Director, ISIS Executive Director 1979–2009

The International Species Information System (ISIS) is an international, non-governmental, non-profit organisation dedicated to serving the worldwide community of zoos, aquariums, government breeding centres and related institutions. It provides cooperatively developed, living collection-management software combined with an increasingly valuable and accurate pool of animal data contributed by all member institutions. Today, it is the world's largest membership organisation of zoos and aquariums, with more than 800 member institutions in 78 countries, involving data on 2.6 million individual living animals and their ancestors.

ISIS began operations in 1974. The founding vision for ISIS came from Ulysses Seal, who began working with zoo animals to do comparative biochemistry and endocrinology. He discovered that most zoos in the 1960s and early 1970s did not have reliable records on identifiable individual animals for most of their collection (excepting perhaps a few 'star' animals). When a veterinarian treated an animal, they commonly did not know the previous veterinary history, age, origin, etc. Ulysses Seal saw both short-term animal-management benefits and long-term conservation and science benefits from a reliable system of identifying individual specimens, and also from pooling veterinary and basic animal data.

Today, the documented individual animal history at most zoos starts from whatever year the institution joined ISIS. There are a handful of zoos with excellent historical specimen records for most of their collection pre-ISIS, but they are the exceptions.

Ulysses Seal was based in North America and had many zoo contacts there. When the call went out in 1973 for zoos to join ISIS, 52 North American and three European zoos joined. Recognising that many zoological animals move internationally and therefore that pedigree and provenance information require global cooperation, an 'I' for 'International' was added as the first letter in ISIS' name. By 1979, ISIS had grown to 85 members and Ulysses Seal moved on to become Chair of the newly reconstituted IUCN/SSC CBSG. Nate Flesness, Ulysses Seal's graduate student, took over ISIS (as half of its total staff at the time, Kim Hastings being the other half). During this time, ISIS provided member institutions with rather unfriendly standard paper animal records forms and keypunched copies of these forms into mainframe computers.

By 1983, several ISIS members had tried to develop mainframe-based, single-institution animal-records systems with limited success. Personal computers were just coming into common use. ISIS proposed to three large institutions that they co-sponsor an ISIS experiment to build a PC-based institutional animal records system designed to be ISIS compatible. The Wildlife Conservation Society, the Brookfield Zoo and the National Zoological Park in Washington, DC, put up US$ 10,000 each. An Institute of Museum and Library Services Grant for an additional US$ 25,000 was obtained, with the promise from ISIS that up to 25 institutions would eventually use the software. ISIS hired its first staff programmer, Paul Scobie, and 12 months later, in late 1984, ISIS delivered ARKS 1.0 (which offered just five functions: consistency-checked data entry, specimen report, taxon report, eight-column inventory report and automated data submission to ISIS).

ARKS 1 was a huge success. The sponsoring institutions dropped their internal software development efforts (into which more than US$ 500,000 had already been invested). More than 100 institutions adopted ARKS 1 and another round of pledges and grants funded development of ARKS 2. ISIS moved from 'just' a central shared animal database to become an economically efficient cooperative software developer for the zoological community. Veterinary (MedARKS) and studbook (SPARKS) software packages were developed after this.

In subsequent years, the Executive Director of ISIS, Nate Flesness, was often invited to annual IUDZG/WZO meetings. This was a great opportunity to meet directors from many institutions and to develop cooperative relationships. In 1989, ISIS incorporated as an independent member-governed, non-profit organisation (US 501c3) under an international Board of Trustees. The Minnesota Zoo, which had administratively 'incubated' ISIS for 15 years, generously supported its becoming fully independent.

In 1994, ISIS was asked by the WZO Council to become their first 'permanent' office on a part-time basis, increasing organisational support and continuity. For the subsequent six years, ISIS Executive Director Nate Flesness and ISIS administrative manager Kim Hastings were the first WZO staff, part-time, under an agreement between the ISIS Board of Trustees and WZO Council. They helped to organise Annual Conferences, edited the conference proceedings, distributed the first newsletter and wrote the minutes for Council sessions. When it was time for an expanded and more open membership, ISIS staff supported Council in lengthy revisions of the WZO Constitution. When a full-time secretariat was envisioned for WZO, ISIS staff helped set up the fund-drive to pay for the first WZO Executive Office in Berne, Switzerland.

Peter Dollinger, the first WAZA Executive Director, in the original WAZA Executive Office established in Berne in 2001 (© WAZA)

Doing part-time work for WZO was a privilege, but ISIS understood from the beginning that it was temporary. In 2000, all WZO office archives and activities were forwarded to WZO President Willie Labuschagne of the National Zoological Gardens in Pretoria. Shortly after that, the first WZO (renamed WAZA) Secretariat was established in Berne.

In addition to ISIS' involvement with WAZA's first permanent office, ISIS and WAZA have partnered to improve community access to animal-management information resources. They have co-sponsored the electronic distribution of studbooks since 1996. Today, these assemblies include about 98% of the known international and regional studbooks of the world.

Chapter 4

The Evolving Role of the Zoo

The 21st century zoo must be redesigned as a hedge against biotic impoverishment; a time machine buying continuance for faltering wildlife populations; a corridor of care between parks and reserves; and, more than ever, humanity's primary introduction to wildlife, promoter of environmental literacy and recruiting centre for conservationists.

William Conway, Past President and General Director of the Wildlife Conservation Society

The Evolution of Zoos

Austrian Emperor Franz Joseph I together with daughter Valerie and four grandchildren at the menagerie in Vienna
(© Vienna Zoo)

A Brief History of Zoos

Collections of wild animals have evolved within the context of their times. Beginning about 5,000 years ago with the earliest urbanised civilisations, Egyptian pharaohs, Assyrian kings, Chinese emperors, Roman emperors and Sicilian kings decorated their palaces and gardens with collections of exotic animals as signs of their power and wealth, and as treasures for their personal enjoyment. Private and privileged collections evolved into public, cultural institutions during the European Renaissance period (14th to 17th century), where a flush of new wild-animal collections spread across Europe as global exploration, economic expansion and intellectual curiosity intensified and gathered speed. Collections of animals kept in cages came to be called menageries during and after the European Renaissance, and were characterised by having as many species as possible exhibited in taxonomically arranged rows of barred cages. The emphasis was on recreation and entertainment, wherein the enclosures were designed for optimal viewing by curious spectators.

The development of large cities in the western world and increasing wealth in the 19th century led to the preservation of natural areas and the design of parkland for recreation. The survival of the natural world became an increasing concern and a demand for a scientific understanding of wildlife developed. During these times menageries evolved into zoological gardens (zoos). Animal displays were living natural-history cabinets that placed animals in cages grouped by their taxonomic classifications in ecological or zoogeographical arrangements. The emphasis was now on education and research.

By the 1930s, many zoo designers in Europe and the USA had begun to think more deeply about the needs of captive animals. Moats and greenery gradually replaced small, sterile, barred cages. Twenty years later, veterinary medicine came of age. More congenial surroundings, better health and good nutrition made for healthier animals. Better conditions for the animals meant better conditions for breeding. By the 1960s, birth announcements dominated zoo news. Faced with an abundance of captive-bred animals and with rising alarm about the possibility of an extinction crisis, zoos suggested breeding rare animal species in captivity as a hedge against extinction in the wild.

Elephants as visitor attractions at the Paris Zoo in 1930 (top) (© Muséum national d'Histoire naturelle) and the Munich Zoo in 1934 (© Munich Zoo)

Zoos across Europe and the USA made research and wildlife conservation through captive breeding their priorities in the late 1970s and early 1980s. By the 1990s and into the new millennium, rapidly increasing numbers of threatened species, limited zoo capacity, lack of habitat for reintroduction and expense made captive breeding as the primary conservation contribution of zoos increasingly impractical. Zoos and aquariums began working to redefine their mission to include a more *in situ*-based approach and embraced the importance of education as a principal role for zoos and aquariums.

An antiquated carnivore house was turned into a visitor conservation centre at the Leipzig Zoo (top), including the display of an outdated lion enclosure (© Leipzig Zoo)

WAZA and the Evolving Role of Zoos

Since its inception, WAZA's members have repeatedly asked themselves what WAZA stands for. They have oscillated between distinguishing WAZA as a conservation organisation, to defining it as an organisation that supports research and education, to expressing that it represents all three aims. Over time, they have also asked themselves how recreation fits into the role of zoos. Generally speaking, WAZA's priorities and development as an organisation since its founding have mirrored the historical evolution of zoos. As most zoos in Europe and the USA in the 1940s, 1950s and 1960s were emphasising the importance of education and research, so too did IUDZG in its earliest days. Notably, IUDZG also stressed the importance of conservation from the onset, acting ahead of its time in this regard. IUDZG's commitment to the core priorities was reflected in Article 2 of the 1947 Constitution:

'The object of the Union is to promote cooperation between directors of Zoological Gardens and Zoological Parks managed on a scientific basis, that is, entirely non-commercial institutions with cultural and educational aims in which the public is allowed to watch and study live animals and which promote zoological research in the widest sense and also the protection of the world's fauna.'

Evolution of Zoos

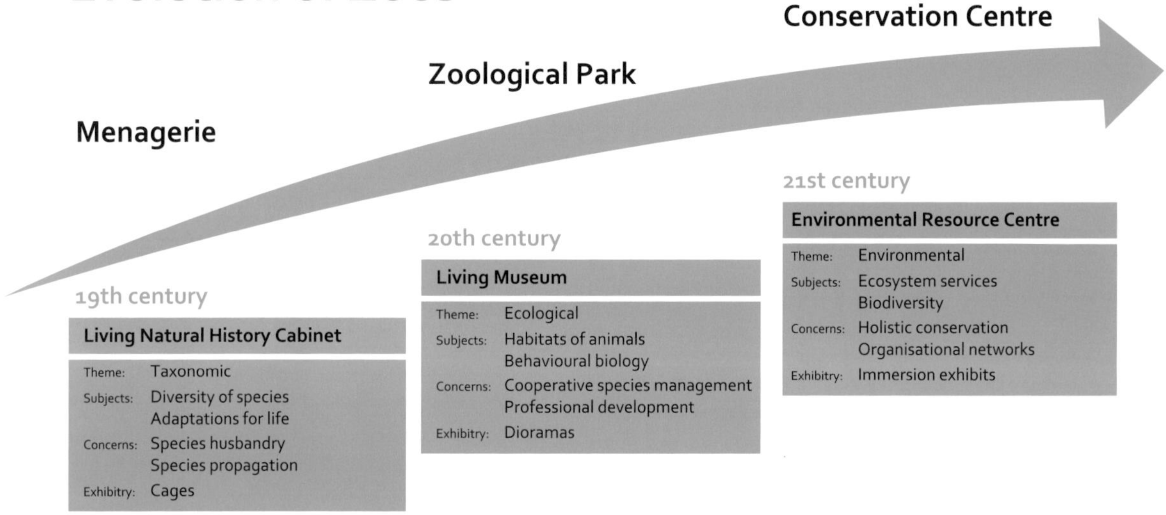

Conservation Centre

Zoological Park

Menagerie

19th century

Living Natural History Cabinet

Theme:	Taxonomic
Subjects:	Diversity of species Adaptations for life
Concerns:	Species husbandry Species propagation
Exhibitry:	Cages

20th century

Living Museum

Theme:	Ecological
Subjects:	Habitats of animals Behavioural biology
Concerns:	Cooperative species management Professional development
Exhibitry:	Dioramas

21st century

Environmental Resource Centre

Theme:	Environmental
Subjects:	Ecosystem services Biodiversity
Concerns:	Holistic conservation Organisational networks
Exhibitry:	Immersion exhibits

The evolution of zoos (modified from *The World Zoo Conservation Strategy*)

IUDZG's first President Armand Sunier echoed the importance of the core aims of the Union and predicted a future course for zoos in 1952 when he stated that:

'Indeed in future all zoos will have more and more to reckon seriously with the facts that:

- A well run zoo is an increasingly necessary, indeed indispensable educational institution of great social importance;
- The problem of saving many species of wild animals from extinction becomes rapidly more urgent.

Therefore, in the near future, the only excuse for the existence of a zoo will be that it fulfils adequately its educational task, which is wholly comparable with that of a museum, and that it contributes to the preservation of species of wild animals.'

Participants the 1978 Annual Conference in Leipzig
(© Jörg Adler)

Responding to the rising alarm about the possibility of an extinction crisis, zoos in the 1960s began to turn their attention to the role that they could play in the conservation of wildlife. In a paper entitled 'Conservation of Nature – A Duty for Zoological Gardens' (Appendix VI), presented at the 1964 Annual Conference in Taronga by IUDZG member Kai Curry-Lindahl, Director of the Nordiska Museet and Skansen, a call to arms was heard:

'To work for conservation is an obligation every serious institution must undertake if it exhibits living animals for museums or recreational purposes. Only by having such an attitude and undertaking such work can the existence of a zoological garden be fully justified.'

Furthermore, Kai Curry-Lindahl was a pioneer among IUDZG members in emphasising the quadruple role that zoos and aquariums should play, stating that:

'The four main functions of a zoological garden or an aquarium should be conservation, research, education, and public recreation. Few other museum institutions other than zoos, have greater opportunities to achieve such a programme, extension of knowledge of nature conservation, particularly of ecological principles, amongst such a vast public. Zoos are especially well-equipped to do this as they are, in the main, open-air museums which can arrange permanent and temporary exhibits in cooperation with living nature itself.

I have put conservation and research as first items in a zoo's programme deliberately because they form the base on which exhibits and other educational exhibits must rely.'

Propagation of corals at the New York Aquarium (© Gerald Dick)

Throughout the late 1970s and early 1980s, IUDZG members made research and wildlife conservation through captive breeding their priority. Dick van Dam, Director of the Rotterdam Zoo, presented a paper at the 1980 Annual Conference in Pretoria called 'The Future of Zoological Gardens', highlighting the point that conservation through captive breeding should be a priority when he stated 'that conservation must receive an increasingly important place in the management of a zoo and that the first step must be an increased birth-rate in the collection'. Research about methods of captive breeding concerning various species was popular at this time and papers reflecting the results of studies were prevalent in the scientific sessions of IUDZG's Annual Conferences. The role of zoos in education and recreation was not emphasised by IUDZG during this period. By the 1990s and into the new millennium, WZO continued to underscore the conservation and research role of zoos and, once again, included education as an essential component.

The Evolving Zoo

By William Conway, Past President and General Director of the Wildlife Conservation Society

The 'zoo' (zoos and aquariums), with its living exhibits, is inherently dynamic and now, as hands-on guardian of increasingly threatened wild animals, it is changing more rapidly than at any time in its long history. In 1945, when I began as a volunteer keeper at the Saint Louis Zoo, it was simply viewed as a kind of museum. Its live subjects, like the art and artefacts of other museums, were to be collected, admired and used to teach. The fragility and limited lifespans of its collections were recognised in steadily improving daily care but not in the scientific challenges of perpetuation. Zoos reflected the tardy environmental perceptions of their times and took little note of the far-reaching alterations occurring outside their gates that would inevitably affect their goals and responsibilities.

At the time, zoos tended to view their collections competitively; an institution's status was more likely to be determined by the number of species it exhibited than by its use of them for education or wildlife preservation. Zoos were an archipelago of territorial entities: islands on which one species after another would, like holiday lights, appear, shine and soon wink out. Compared with today, most curatorial staff were poorly prepared and veterinarians were often employed only part-time. The lack of significant scientific staff meant that research was limited and, in 1945, fewer than 24 keepers, curators or directors in US and European zoos were women. However, public perceptions and zoos were rapidly changing.

In the late 1950s, educational programmes in zoos and animal-exhibit interpretation leaped ahead. Zoo collections became the subjects of serious science, exemplified by the behavioural studies of Zurich Zoo's Heini Hediger. By the 1960s, advancing curatorial and veterinary sciences made possible the removal of more and more tropical zoo animals from the sterile bathroom-like enclosures once thought necessary for their health. The arts of environmentally sensitive zoo exposition joined its sciences and 'habitat' approaches (pioneered in some zoos much earlier) began to immerse visitors in simulations of the animal's natural habitat.

In the 1970s, the first national zoo accreditation plans and policies were worked out in the USA, soon effecting major changes in zoos and their objectives. Although most zoo-animal breeding efforts at the time ignored population biology, by the 1970s zoo-based genetic studies, especially at the National Zoological Park in Washington, DC, were helping to build a foundation for the new science of 'conservation biology' and the 1980s proved to be a period of remarkable development. There was a flowering of zoo science, care technology and policy collaboration. The (American) Association of Zoos and Aquariums (AZA) created its Species Survival Plan (SSP) for zoo-animal propagation, which was rapidly adopted by the European Endangered Species Programme (EEP). Meanwhile, IUDZG worked to overcome historical limitations as it sought to transform itself from an exclusive group of zoo and aquarium directors to an inclusive world zoo and aquarium association of zoo professionals, and to play a larger role in international conservation.

By the 1990s, it was apparent that protected parks and reserves were becoming more isolated, their wildlife populations smaller and more fragmented, in a word, more zoo-like. At the same time, there were growing numbers of zoo-based *in situ* conservation programmes being implemented, making zoos among the largest of all non-governmental sources for international wildlife conservation support, now annually exceeding US$ 350 million. But wildlife is continuing to disappear.

In 2010, the Living Planet Index announced that there had been a 30% decline in wildlife populations since 1970, adding that more than 47,600 species are threatened by extinction. The population of even such iconic species as lions has dropped 90% in the last 70 years and less than 4,000 tigers remain in nature. In 1945, the human population was 2.5 billion. It will reach seven billion in 2011. People now influence over 83% of the Earth.

Future zoos must not only meet the needs of the wild animals they care for but also must become reservoirs of rare wildlife and centres for its support in parks and reserves. The survival of protected wild animals is becoming the zoos' ultimate education and conservation goal.

Lion (© Nicole Gusset-Burgener)

Conservation and the Future of Zoos

By Jenny Gray, CEO of Zoos Victoria

Imagine holding the last Tasmanian devil in the world in your hands, you smell his musky odour and you see his once shiny nose, now cracked and dull. His once pure black pelt is grey with age, his limbs are stiff. Imagine it is your job to euthanise the last devil in the world. Imagine the sadness that will fill your soul when he is gone. We can never let this happen.

Zoos and aquariums have enormous power and resources. With power comes great responsibility. We have chosen to hold and care for the animals of the planet, thus it is our duty to ensure that they survive. Every day millions of people choose to visit zoos and aquariums to see the wondrous animals that share our planet. Our desire to learn about animals is insatiable, our affection immense.

Within zoos and aquariums we allow our minds to journey to the wild places of the world. We marvel at the adaptations that allow life to thrive at every extreme habitat, we gasp in wonder at the beauty that surrounds us. But despite the awe and wonder we know that the world is not as it should be, many species are under a threat that has the potential to destroy them forever. The billions of people that now inhabit the Earth are destroying the habitat and environments that animals need to survive. We are on a trajectory that will see the diversity of Earth shattered and we will not be able to return, nothing we can do will ever bring back extinct species.

The power of zoos and aquariums needs to be turned to all aspects of conservation, preserving threatened species in human care, developing safe places for release, breeding significant and diverse populations and, most importantly, working with partners, governments and communities to protect the remaining wild places. Zoos and aquariums are advocates for the animals of the world. By bringing animals into the hearts and minds of people, zoos and aquariums create bonds between people and animals, bonds that remind us to act in ways that help animals. As conservation organisations zoos and aquariums are uniquely placed to empower people to act differently, to minimise the harms that are inflicted upon the animals that share our planet.

Our animals and our staff are the most engaging and powerful reminder that people need to change. We need to forge a new pact with animals, a pact that recognises that animals are important and compels us to act in their interest, even at the cost of our own. If we do all that we can, it may just be that when we hold an old devil in our hands, we will feel joy at his life, we will see his descendants growling and fighting over a rotting carcass. We will be able to celebrate that we helped this tough little beast overcome great odds, facial tumours, habitat destruction and introduced predators to remain the top carnivore in Tasmania.

Tasmanian devil (© Christian Schmidt)

The Anti-Zoo Movement

IUDZG members first described that an anti-zoo movement was emerging in 1984. Around this time, more and more aquariums had become involved in displaying dolphins, resulting in a build-up of militant opposition to maintaining cetaceans in captivity. Discussions among members about how to confront the increasingly charged issue were started at Annual Conferences during this period. By 1986, IUDZG wrote a statement on cetaceans in order to indicate their position on the subject for the benefit of their associates, the public and their critics:

'The IUDZG meeting in Wroclaw expressed concern at the proliferation worldwide of new, but differing, standards in respect of Cetacea in human care.

The Union welcomes any proposals which seek to improve the husbandry of such animals, but it can only support those rules and regulations that are based on experience and scientific fact.

The Union also supports the use of Cetacea for research, education and other conservation roles.

The Union agrees to support the international, national and regional areas with IUDZG membership in speaking with the legislators of such regions when proposals do not have scientific validity.

The Union can not take up individual cases in this respect.'

Dolphins and Fisheries – A Modern Dilemma

By Shigeyuki Yamamoto,
Chair of the Japanese Association of Zoos and Aquariums (JAZA)

WAZA is a global association of zoos and aquariums that play an important role in biodiversity conservation, environmental education, biological research and as social centres, connecting people with vanishing nature. It is WAZA's primary goal to reduce the magnitude of the current extinction crisis and one of the strategies under that goal is to translate developing knowledge about animal ecology, behaviour and health into standards of care that can be promulgated across a very diverse membership, a diversity of culture, wealth, laws and traditions. WAZA respects all of its constituent cultures and recognises the need for sensitivity and constructive engagement.

WAZA members take great responsibility in caring for marine mammals, as they serve as public ambassadors for marine conservation. The connections created between visitors and these remarkable animals play a key role in providing the support, research and public awareness needed to conserve and protect the marine environment. Additionally, many species are vulnerable to extinction and developing self-sustaining captive populations may act as a reservoir against extinction. Had there been a concerted and timely use of a captive-breeding insurance policy for the Yangtze River dolphin for example, we might not now be mourning the recent extinction of this species.

The dolphin drive fishery in Japan has generated a lot of fierce debate and public outcry, and WAZA has played an active role in mediating between opposing factions. The whaling industry in Japan has been managed by the Japanese government and has a long history and established cultural significance. A by-product of this industry is the collection of dolphins for public aquariums in Japan and for other countries. The method of capturing these animals has caused great concern to WAZA in terms of animal welfare and WAZA has maintained an approach of constructive engagement with members in this regard, observing that more militant action has met with little positive response.

Bottlenose dolphin
(© Gerald Dick)

WAZA has worked with JAZA to establish a new, transitional approach to this component of the inshore fishery by looking to separate the 'fishing' or harvesting from the collection for aquariums, by dramatically improving the methods used both to address the welfare concerns and by developing population-management programmes to allow aquarium animal populations to become self-sustaining. As population sustainability increases, the need to obtain additional founder animals through wild capture decreases.

JAZA has been a vital partner in this progress so far and continues to encourage technical exchanges of information and advanced genetic management. There is a platform for discussion and development of new ideas and approaches because of the WAZA–JAZA relationship. This is a very powerful way to work with Japanese colleagues. The tradition stretching back generations is being reviewed within Japanese society. Considerable progress has been made and the discussions continue.

A Review of Elephant Management in European Zoos

By Stephan Hering-Hagenbeck, Director of the Hagenbeck Animal Park

Humans have kept elephants, especially the Asian species, as working, war, show and temple animals over centuries. However, humans have never been able to domesticate the grey giants. Elephants are one of the most recognised and important species kept in zoos. They are the best example of a flagship species.

When these wonderful and impressive creatures were introduced into the European zoo world more than two centuries ago, they were managed according to traditional management techniques, which were copied from the different Asian cultures where the animals originated. Elephants, which in those days nearly all came out of their country of origin, were either received 'broken' as in the case of all the Asian elephants or they were tamed with the same brutal procedures, as learned from Asian mahouts.

Some animals were even imported together with their mahouts. Generally, due to lack of knowledge, most of the animals were not really managed at all. This often resulted and still results in major foot problems that frequently ended in losing the animal as a result of secondary infections.

Leading European zoos quickly learned that elephants in captivity needed a lot of attention and care from highly qualified staff. These facilities slowly developed their own ways of training and managing the different elephant species. Different attempts resulted in highly sophisticated 'hands-on' and later 'protected contact' management programmes, which all had one major factor in common: they removed the need to 'break' the elephant!

Although we, as the zoo community, still need to improve many things in elephant husbandry, management, health care, *ex situ* breeding and facility design, it can be stated that leading European zoos have developed their own way of keeping these animals in captivity. These husbandry methods are currently the most successful elephant-management and breeding programmes in the world. However, we cannot rest on our laurels. There are still major goals that need to be achieved in the near future, such as how to treat still fatal diseases, such as herpes and tuberculosis, how to increase our breeding success, how to solve the surplus male problem and how to create better, more enriching and nature-like habitats in order to improve the social day-to-day lives of the animals.

African elephant (left) (© Nicole Gusset-Burgener) and Asian elephant (© Gerald Dick)

As zoo professionals, we should always be aware of the fact that we are only temporary visitors to our facilities, whereas the animals are permanent. They live in habitats that we have created for them 24 hours a day, 365 days a year.

We are only able to spread our conservation message if we manage to portray our animals as active, healthy, reproductive and as natural as possible. They are the ambassadors for their relatives in the wild.

The Asian elephant is potentially one of the best species to spread our message for conservation; we just have to provide these animals with the best possible management and most ideal environment to do so.

Great Apes in Zoos

By Colleen McCann,
Curator of Mammals at the Bronx Zoo/Wildlife Conservation Society

As zoo professionals, we believe that achieving the highest standards of health and husbandry for great apes fulfils their physical, psychological and social needs. Our cooperative breeding programmes enable us to maintain demographically and genetically robust populations in quality zoological environments. This translates into not only ensuring the well-being of individual animals but also is a commitment to the long-term survival of great ape populations and a critical hedge against extinction. Not everyone agrees. There are some outside of our professional community who believe that great apes and other animals should only survive in the wild.

We believe that robust wild populations are essential to biodiversity but also feel strongly that managing assurance populations in our parks is a critically important component to our overall long-term conservation strategy. Others believe that animals should receive the same rights as people. While we strongly believe in the well-being of all animals, imbuing individual rights may very well place the needs of one above the needs of the many and conflicts with our goal of maintaining biodiversity through species conservation. In addition to our commitment to conservation through intensive species management, a diversity of wildlife in zoological parks plays a critical role in education and public awareness regarding the conservation of species and the many threats to wild habitats. In 2005, for the first time in recorded history, the majority of humans lived in urban areas.

The sad truth is that only a very small percentage of people will ever have the opportunity to see great apes in the wild. We believe that the human experience will be greatly diminished if we fail to maintain this connection to wild nature through experiences in our zoos.

Although today's best practices in great ape husbandry have led to the notable success of captive-breeding programmes globally, this comes as the result of decades-long efforts and important milestones in ape-husbandry knowledge. The western lowland gorilla was first known to scientists in the late 1800s. Twelve or so infant gorillas were brought from Africa to Europe and North America to survive only a few days or months at best.

Over the next several decades, gorilla longevity in zoos continued to increase and in 1956 an important milestone was reached when the first gorilla was born in captivity in North America at the Columbus Zoo. In the ensuing decades information on the natural history of gorillas became available from field studies and were applied to our husbandry practices.

In 1959, the Wildlife Conservation Society's (WCS) George Schaller conducted the first study of gorillas in the wild, and provided critical information on their diet and social behaviour. As these findings were incorporated into husbandry practices, zoo populations experienced a marked improvement in their health and reproductive success. The inception of the SSP for gorillas in 1980 ensured a collaborative programme for sustaining a genetically and demographically viable population among North American zoos. A survey conducted in the mid-1970s revealed that only 10% of gorilla infants born in zoos were successfully mother-reared.

We now know the key to having success with mother-reared offspring is allowing females the opportunity to observe and learn maternal skills while being raised in family groups. Today, the AZA Gorilla SSP population has achieved 76% infant survivorship with 85% of infants mother-reared. Gorillas in zoos can reproduce up to the age of 42 years, have a median life expectancy of 29.9 years for males and 36.1 years for females with a maximum longevity record of 55 years, exceeding the observed estimates for gorillas in the wild.

The Bronx Zoo manages its gorilla population in the award winning Congo Gorilla Forest, a multi-species exhibit that features gorillas and a variety of other central African species. This unique immersive wildlife experience transports people from New York to the wilds of the Congo Basin, highlighting the critical work of WCS, and provides visitors with the opportunity to direct their entry fee to an on-going landscape conservation programme in this region of Africa.

The success of this exhibit is evident in a number of ways. Since 1999, over US$ 10 million has been raised for conservation and more than 100,000 students and nearly 2,400 teachers have benefited from education programmes. In addition to 14 gorilla births, we have also achieved success with a number of species including the births of 23 red river hogs, 11 Wolf's guenons and five okapis since the exhibit was opened. This has proven to be a most valuable synergy between our commitment to species in our zoological park and to their future in the wild.

A view into the Bronx Zoo's Congo Gorilla Forest exhibit. Featured are some members of one of the two gorilla family groups, each of which has an expansive outdoor exhibit area (© Julie Larsen Maher/WCS)

Participants at the 1987 Annual Conference in Bristol (© WAZA)

In order to further offset some of the negative criticism affecting zoos and aquariums, IUDZG members suggested that they should collectively emphasise how zoos serve as important resources for education. Charles Hoessle, Director of the Saint Louis Zoo, expressed this viewpoint particularly well at the 1987 Annual Conference in Bristol: 'We must become more prominent in our role in conservation education. Teaching our visitors, our public and our school children about zoo animals and animals in the wild, conservation education, that is our best defence'.

The subject of the anti-zoo movement took up increasingly more time and generated more discussion with each successive Annual Conference. By 1987, IUDZG members had started talking about incorporating a Code of Ethics modelled after the existing code of AAZPA. However, members such as Frederic Daman, Director of the Antwerp Zoo, cautioned his colleagues, mentioning that it was 'perhaps too soon to have an international code of ethics and that we should remember that a large part of the zoo world does not have an organised zoo association, leading to inconsistencies'.

Arabian oryx (© Christian Schmidt)

By 1988, in addition to highlighting the educational benefits of zoos, members discussed publicising the important conservation role of zoos by drawing attention to their positive stories about conservation successes locally, nationally and internationally. They considered emphasising the fact that many new births had occurred as a result of inter-zoo cooperation and research and they discussed the possibility of sharing the results of reintroduction programmes for animals such as the Arabian oryx, golden lion tamarin and Bali mynah. Some members suggested releasing the statistics from ISIS that indicated that a larger portion of mammals in US zoos were captive-born. In the same year, Charles Hoessle gave a report indicating that 'most of the anti-zoo campaigns distort the truth and attempt to organise letter-writing campaigns to generate bad press, regardless of the facts. The facts are not important to them and because of this, confrontations are not desirable by the zoos and not productive'. Furthermore, Charles Hoessle stated that 'zoos represent the real animal rights people, our goals are species oriented and not individual animal oriented. Our goals are long term'. Roger Wheater, Director of the Edinburgh Zoo, suggested that 'what we have to do is develop an IUDZG strategy for dealing with this matter. It is important to have as much ammunition as possible, either to be reactive or more importantly, pro-active to this particular problem, which is not going to go away and will face us more and more in the years to come'.

Bali mynah (© Christian Schmidt)

Despite the charged rhetoric of the previous year, by 1989 members decided to put the topic of the anti-zoo movement aside in order to go on with 'really important issues'. (Discussions about the anti-zoo movement did not formally end in 1989; further discussions on the subject came up in other years and specific examples of ethics violations (of non-WZO zoos) were also frequently a subject of active discussion.) In 1990, a working group was formed in order to look into international communications and the implications for the Constitution and structure of the organisation. They determined that the highest priority among the immediate needs for action for IUDZG was to create a Code of Ethics.

WAZA's Code of Ethics and Animal Welfare

Members of IUDZG supported the cause of animal welfare early on. At the 1956 Annual Conference in Chicago, members declared that IUDZG 'completely endorses the efforts of [the Royal Society for the Prevention of Cruelty to Animals (RSPCA)] and that it will support and uphold its principles'. IUDZG's motion and resolution stated that:

'Whereas, it is known and deplored by most men and should be known by all men that certain practices and exorbitant licence exist and are allowed which would deprive animals of their inherent right to live in their countries of origin and habitats therein, and in such other proper places elsewhere in the world as they might be kept and propagated and thereby survive,

BE IT THEREFORE RESOLVED, that the International Union of Directors of Zoological Gardens in meeting on this day, the eighth of June 1956, do now and thereafter condemn these practices,

Participants at the 1956 Annual Conference in Chicago (© WAZA)

BE IT FURTHER RESOLVED, that the International Union of Directors of Zoological Gardens will by every fair and proper means exert and implement every effort to bring such trade and practices to an end, and to create a fair will of all men towards animals.'

The text of the resolution was telegraphed to the executives at the Congress for Conservation of Nature held in Edinburgh, UK, in 1956 to underscore IUDZG's stance. Many years passed until the Union would concentrate on the theme of animal welfare again. It was not until 1993, when a combination of the residual momentum from previous years of the anti-zoo movement and the publication of a book called *The Great Ape Project – Quality Beyond Humanity* edited by Paola Cavalieri and Peter Singer, refocused WZO's attention on the subject once more.

Orangutan with child at the Vienna Zoo in 1971 (© Vienna Zoo)

Members participated in a lengthy discussion about animal welfare and the book's suggestion that great apes should be given equal rights with humans. WZO's first working group on ethics was established in 1994. During its first year of operation, the working group discussed the problem of WZO being a Union of individuals and associations that represent different cultures and beliefs, thus having varying views on the subject of animal welfare. WZO President Gunther Nogge, Director of the Cologne Zoo, elucidated this point by stating that 'it is a fact that there is a great variety among the hundreds of zoological institutions worldwide which are now gathered as members under the umbrella of the world zoo organisation. The differences in zoos mirror their different cultural, historical, political and economic backgrounds. Regardless of these differences, all zoos, wherever they may be, must provide their animals with the best possible living conditions according to our best knowledge'.

Originally, the working group on ethics recommended that WZO mandates or encourages its regional associations to adopt their own ethics codes, which would reflect their individual cultures. Acknowledging the stated cultural differences between zoos, however, the working group abandoned its recommendation, feeling that such a strategy would lead to confusion and, therefore, weaken the impact of a WZO ethics code. Instead, the working group's unanimous approach was to draft a statement on ethics that would set forth, in a narrative format, a set of ethical guidelines which would broadly define how members work with each other and with the animals under their care. Following this narrative, the working group would list a set of mandatory standards, which would define more specifically those ethical standards of greatest concern, and would base their template on the EAZA and AZA Code of Ethics. The first draft *Code of Ethics* was submitted in 1997, a second draft was submitted in 1998 and by 1999, the Code was formally adopted at the Annual Conference in Pretoria.

A sub-group called the welfare working group was established alongside the working group on ethics in 1996. It was formed in order to progress WZO's priority of maintaining high standards in animal husbandry and veterinary care and to help less-developed zoos. One of its main goals was to prepare a database on animal welfare, cataloguing publications, training opportunities and details of experts on the subject for the benefit of the WZO membership. From 1997 to 1999 it worked to this end. By 2000, there was no further mention of the welfare working group; instead attention was focused on the newly formed Ethics Committee, chaired by Ed McAlister, Director of the Adelaide Zoo, which was established to deal with complaints about ethics-related violations.

The Ethics Committee produced a unanimously adopted *Code of Animal Welfare* in 2002. The same year, the Committee changed its name to the Ethics and Welfare Committee. By 2003, the Code of Ethics and the Code of Animal Welfare were combined and re-written to form a joint WAZA *Code of Ethics and Animal Welfare*, which was unanimously accepted by the WZO membership (Appendix II). The next priority on the agenda for the Ethics and Welfare Committee was to prepare a Code of Research. The code, entitled *Ethical Guidelines for the Conduct of Research on Animals by Zoos and Aquariums*, was adopted in 2006.

Polar bear (© Rick Barongi)

Environmental Education

'Performing' animal keepers at the Paris Zoo in 1934 (© Muséum national d'Histoire naturelle)

A Brief Overview of Education in Zoos

Today's zoos provide a plethora of educational opportunities that are designed to engage, stimulate and inspire visitors. From interactive graphics at exhibits to puppet shows, zoo education attempts to connect visitors to wildlife and wild places. Zoo education has evolved significantly since the birth of zoos. In the earliest days, there was no emphasis on education at all. Living animals were on show and entertainment was the only objective. The 19th century was an age of discovery in zoos across Europe and the USA. While the main interest was in entertainment, consisting of such things as performing animals and animal rides, taxonomy was also beginning to be a focus. The early 20th century saw the beginnings of interpretation in the form of basic animal labels, zoo-keeper talks and feeding shows. By the 1950s, there was an emergence of 'education' as an employment track in zoos. Schools requested educational activities, such as tours and classes, increasingly using zoos as part of the curriculum. Themes of adaptation, behaviour and taxonomy were explored. The 1980s saw the widespread employment of zoo educators and focused on cross-curricular themes, and on creating and disseminating educational 'packs' and resources. In the 1990s, zoo education grew to focus on conservation themes, including public education about human–environment relationships. Although these subjects are still an active part of the education themes in zoos, today's zoo-education programmes include a more substantial focus on inspiring action and behavioural change among zoo visitors in favour of protecting wildlife and wild places.

'Performing' animal keepers at the Vienna Zoo in 1910 (© Vienna Zoo)

Zoo education is based on the principles of environmental education. The classic definition of this term was formulated in 1970, in a joint effort between IUCN and the United Nations Educational, Scientific and Cultural Organisation (UNESCO):

'Environmental education is the process of recognising values and clarifying concepts in order to develop skills and attitudes necessary to understand and appreciate the interrelatedness among man, his culture, and his biophysical surroundings. Environmental education also entails practise in decision-making and self-formulation of a code of behaviour about issues concerning environmental quality.'

Learning about gorillas at the Cologne Zoo
(© Cologne Zoo)

The goals and objectives of environmental education were devised later in 1975, at an UNESCO meeting in Tbilisi, Georgia. Called the Tbilisi Declaration, the five objectives for environmental education are to instil:

- Awareness and sensitivity to the environment and environmental challenges
- Knowledge and understanding of the environment and environmental challenges
- Attitudes of concern for the environment and environmental challenges
- Skills to identify and help resolve environmental challenges
- Participation in activities that lead to the resolution of environmental challenges

WAZA and Education

The first paper on an educational subject at an Annual Conference was on 'Signs Made of Glass in Zoological Gardens' in Rotterdam in 1957. A second educational paper 'On Labelling in Zoological Parks' was presented at the 1963 Annual Conference in Chester. In 1968 in Pretoria, a paper was presented on the subject of 'Conservation Teaching' in zoos. In 1972 in Amsterdam, eight papers were presented at the Annual Conference and the conference itself was the first one to focus on 'Education in Zoos'. In 1974, the Union was approached by the World Wildlife Fund (WWF, now the World Wide Fund for Nature) and offered the opportunity to set up a joint project for conservation education. In the same year, the International Union of Education Officers made an inquiry about sending their observers to the Annual Conference. Both initiatives were rejected by IUDZG.

IUDZG was slow to respond to the role of education in zoos. Throughout the 1970s and mid-1980s, IUDZG members felt uneasy about the intervention of outside international bodies, such as the International Association of Zoo Educators (now the International Zoo Educators' Association, IZE). Members wanted total control of their collections, zoo-based programming and internal policies and did not want to have their authority challenged 'by their subordinates'. In 1985, for instance, the Council discussed whether the President of IZE should be invited to IUDZG's Annual Conference and they 'agreed that such a move might be premature and might disturb some colleagues'. Their close-minded attitudes reflected an 'old-fashioned' approach to zoo management, wherein zoo directors had full power over all zoo matters. It was not until 1988, when the momentum of zoo-based educational programmes caught up with the members of the Union, that IUDZG drafted its first policy on education. It read:

'The Union recognises that, if conservation strategies are to succeed, the society we live in must be supportive of these strategies and also support will be more forthcoming if society has an understanding of the issues involved.

Zoological institutions represent an educational resource, which can make an important contribution toward this understanding by creating a heightened awareness of these conservation issues, and where appropriate, the solutions. It is the duty of members to endeavour to ensure that their resources be used to their full potential for this purpose.

The social and cultural functions of zoological institutions are, to a high degree, achieved through educational activities. Members should therefore develop and sustain multifaceted programmes of education on the living world, particularly ecology and the need to conserve plant and animal life and how this may be achieved.

Bat education programme in India
(© Zoo Outreach Organisation)

Environmental education at the zoo
(© National Zoological Gardens of South Africa)

Zoological institutions have an important role to play in the implementation of wildlife conservation strategies and members are thus urged to publicise this role as a part of their educational programme.'

In 1994, IZE was officially introduced at that year's Annual Conference in São Paulo. By 1995, the Union sought to actively cooperate with IZE and, therefore, setup a working group to strengthen linkages between the two organisations. Additionally, it sought to gain support from IZE in implementing the WZCS, which had been published in 1993. By 1996, IZE was formally selected as the:

'professional organisation charged with the responsibility for integrating conservation education into the WZCS'. The Union resolved that 'education is to be considered henceforth, one of the primary tools for conservation and the Union will exercise leadership in promoting this concept to its member institutions and provide the necessary support for its implementation'.

In 2002, IZE became an affiliate member of WAZA and was given the distinction of being WAZA's 'education arm'. Despite this extension of responsibility for IZE, there was a feeling in the same year of 'a lack of clarity about the linkage between IZE and WAZA'. As a result of this sentiment, the question was raised about how IZE wanted to be represented and whether WAZA should have a separate Education Committee. Annual Conference members stated that:

'Education should not be an issue only for IZE, but ought to have a specific role in the WAZA Council. Therefore, it should be represented as a separate group. The education representation within WAZA should be a WAZA Education Committee with strong representation from IZE and also directors on the Board.'

Thus a new Committee was established by the Council and Henning Julin, Director of the Aalborg Zoo, was suggested as Chair. Henning Julin attended the IZE conference in 2004, illustrating a growing cooperation between the two organisations. Shortly after, it was decided that WAZA would host IZE on their premises in Berne, Switzerland, supplying office space and administrative support. The year 2007 marked a further milestone, with the acceptance of an official charter for the WAZA/IZE Education Committee, delineating the responsibilities of each organisation towards fulfilling the goals of conservation education. Since the historic charter, WAZA and IZE have worked together on two joint projects, the publication of a gorilla education manual in 2009 and the publication of an education manual called *Biodiversity is Life* in 2010. The first joint WAZA/IZE conference was held in Adelaide in 2008. In 2011, WAZA and IZE signed an MoU with the objective of providing guidance in identifying specific projects and other forms of collaboration between the Parties. Also in 2011, a decision was made to house all future administrative duties in the IZE President's office, removing WAZA's hosting function.

Mountain gorilla, Uganda (© Gerald Dick)

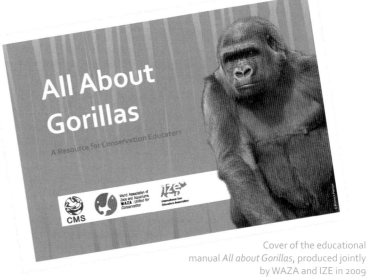

Cover of the educational manual *All about Gorillas*, produced jointly by WAZA and IZE in 2009

WAZA and the International Zoo Educators' Association (IZE)

By Stephen McKeown, IZE President 2006–2010
and Head of Discovery and Learning at the Chester Zoo

IZE was established by a group of European zoo educators in Germany in 1972, 37 years after the birth of IUDZG. One of its first declared aims was to hold a conference every two years, which to date it has, and from 1977 when the first newsletter was published, to publish a journal or newsletter at least annually. This schedule too was adhered to with the exception of two gaps in 1991 and 1998. These publications are an interesting record of the evolution of zoo education over four decades. In that first journal published in 1977, the first President of IZE, Han Resenbrink of the Amsterdam Zoo, noted that 'nowadays a zoo often employs an education officer or even an education department. Zoo education is a comparatively new field and many of the activities stem from the enthusiasm and idealism of the pioneers'.

Forward-thinking zoo directors had already recognised the importance of zoo education some years earlier. At the 1964 symposium 'Zoos and Conservation' held at the Zoological Society of London, and sponsored by IUCN and IUDZG, George Mottershead, Director of the Chester Zoo and IUDZG President, initiated and chaired a session on 'Conservation Education in Zoos', one of the first of its kind.

The educators in IZE increasingly realised that if zoo education was to evolve in a meaningful way it was vital to gain the support of the zoo-director community. To that end, at IUDZG's 1994 Annual Conference, IZE was officially introduced and a report given on behalf of the IZE President. During the following year a working group was formed to look at how the two organisations could work together more effectively and in 1996 IZE was formally charged with the responsibility of integrating conservation education into the *World Zoo Conservation Strategy*.

In doing this IUDZG resolved that 'education is to be considered henceforth, one of the primary tools for conservation'. This led to IZE overseeing the writing of the education chapter of WAZA's 2005 publication, *Building a Future for Wildlife: The World Zoo and Aquarium Conservation Strategy*.

In 2000, the year in which WZO was rebranded as WAZA, discussions were had about overlapping an IZE conference with that of WAZA. This became a reality in August 2002 at the Vienna Zoo when the end of the WAZA Annual Conference overlapped with the start of the IZE meeting. So, in the opulent surroundings of the Emperor's Pavilion, the zoo directors of WAZA and educators of IZE mixed together and started to further formalise the relationship between the two organisations over cocktails.

In fact, a resolution had been passed by WAZA at that 2002 meeting, recognising IZE as its official education arm and IZE Board members were invited to become members of the WAZA Education Committee. Later the IZE President would become co-Chair of the Committee in an effort to further strengthen relations.

However, with a growing membership and an increased level of activity, IZE faced various logistical issues in that it had no formal base or secretariat, and functions such as banking and journal production moved as Board members changed. In 2004, IZE President Chris Peters negotiated a deal with WAZA whereby IZE was invited to establish a central office at the WAZA Secretariat in Berne, Switzerland. This came with the generous offer of the equivalent of one day per week administrative support. This worked extremely well and both organisations gained a greater understanding of the other's work in this new relationship.

However, financial realities meant that this function had to become a charged-for service from 2009 and, after conducting a review, the IZE Board decided in 2010 that the bulk of the administrative function, mainly accounting and membership, would move to the IZE President together with a budget for this. An MoU between IZE and WAZA was written in 2010 and signed in October of that year by the WAZA Executive Director. In 2011, an offer was gratefully accepted to accommodate residual IZE administrative functions at the Secretariat for the European Association of Zoo and Wildlife Veterinarians in Berne, which also serves as an official address for IZE.

The Heini Hediger Award

In 1996, WZO established an award called the 'Heini Hediger Award' to honour those who have 'dedicated their careers to the betterment of zoo professionals and the animals they cared for'. The award was named in honour of the founder of modern zoo biology and the long-time Director of the Berne Animal Park, Basel Zoo and Zurich Zoo, Heini Hediger.

Since the establishment of the Heini Hediger Award in 1996, 12 people have received the distinguished prize.

Recipients of the Heini Hediger Award from 1996 to 2011

Year	Awardee
1996	Ulysses Seal
1996	George Rabb
1998	Jeremy Mallinson
1999	William Conway
2001	Roger Wheater
2004	Sally Walker
2006	Gunther Nogge
2007	Willie Labuschagne
2008	Peter Dollinger
2009	Karen Sausman
2010	Leobert de Boer
2011	Gordon McGregor Reid

Leobert de Boer received the
Heini Hediger Award at the 2010 Annual Conference in Cologne (top) (© WAZA)
The Heini Hediger Award is a hand-crafted ceramic figurine from Uruguay (© WAZA)

Heini Hediger (1908–1992) – A Short Biography

By Alex Rübel, WAZA President 2002–2003 and Director of the Zurich Zoo

Few people in the 20th century have contributed as much to the development of zoological gardens in the world as Heini Hediger. As a zoologist and animal psychologist he is held to be the founder of modern zoo biology. He was successively director of three Swiss zoos in Berne, Basel and Zurich.

His development as animal psychologist, zoo director, scientist and academic teacher began in his parents' garden in Basel. Young as he was at 12 years old and wearing a cap marked 'Zoo', he built his own zoo and undertook his first observations of animals. He studied zoology at the University of Basel and was given the opportunity of accompanying an ethnographic expedition to Melanesia in order to collect amphibians, reptiles, birds and small mammals for the Basel Natural History Museum.

Even at this stage, he was concerned not only with simply collecting material, but also with getting to know the territories and behavioural patterns of the various animal species. He was also generally concerned with contributing to the understanding of relationships between animals and their environment. His herpetological recordings were among the first in the field and contained notes on the life histories of the reptiles he collected. The results of his observations provided him with the basis for a comprehensive thesis for which he received a Doctorate of Zoology at the University of Basel. His results also led to the description of three new taxa, which were named after him. His profound knowledge of herpetology led to his appointment as Curator of Reptiles at the Basel Natural History Museum in 1937.

But Heini Hediger veered towards a field of science that was then still in its infancy – animal psychology. At that time in Germany and in the USA, animal psychology was carried out in an uncritical manner in two different directions ('the machine theory' and the 'humanisation' of animals) and was not taken seriously in scientific circles. Heini Hediger rejected both theories. His starting point for understanding animals was to undertake detailed research into their characteristic behavioural patterns in the wild. His discoveries were based on detailed observations in zoos (Hagenbeck, Frankfurt and Marseille) and circuses, in which the relationship between animals and human beings played a vital role.

Cover of the book *Man and Animal in the Zoo*, published in 1965

Cultural prize of the city of Zurich: Nobel Prize winner Konrad Lorenz (left) and Heini Hediger at the ceremony in 1974 (© Zurich Zoo)

A childhood dream was fulfilled when Heini Hediger took over the leadership of the Berne Animal Park in 1938. He designed the enclosures from the perspective of the animals concerned and with consideration for human-animal relationships. He managed to demonstrate that with optimal space and appropriate husbandry, even the breeding of difficult animals such as the European hare could succeed. During this period he published his results in his book *Wild Animals in Captivity: An Outline of the Biology of Zoological Gardens,* with which he revolutionised the keeping of animals in zoos and founded the science of zoo biology.

In 1944, Heini Hediger took over the management of the Basel Zoo. He began discussing zoo-biology issues at symposiums all over the world. In 1952, in recognition of his groundbreaking work in the field of behavioural research describing the 'flight distance, the individual distance, the social distance and the critical distance', he was awarded an Honorary Doctorate by the Veterinary Medicine Faculty of the University of Zurich.

After disagreements with the Board members of the Basel Zoo he left to become Director of the Zurich Zoo in 1954, delighted to have full rein to practice his zoo-biology principles in the Africa House, replete with tropical plants. For the first time, he was in a position to show the general public the symbiosis of hippopotami and rhinoceros with oxpeckers in a heavily planted environment without fixed walls, right angles and flat surfaces.

His practical experiences of zoo biology were published in 1965 in *Man and Animal in the Zoo*. All of Heini Hediger's publications on animal psychology aimed to ensure a better understanding of animals and to provide many new insights. It was a matter of great importance for Heini Hediger to impart his knowledge of 'primitive zoology', as he called it, to others and his lectures on the radio and his own early TV shows, where he usually brought some animals to the studio to comment on them, were appreciated. As joint publisher and editor with Konrad Lorenz and Bernhard Grzimek of the animal magazine *Das Tier*, and as author of numerous scientific works in the field of zoo biology and animal psychology, he was recognised as an outstanding zoo director the world over.

Heini Hediger was a founding member of IUDZG in 1946 and became its President in 1949, although his strengths and knowledge were in zoo biology more than in managing the Union. His advice was called upon for the planning of zoos in São Paulo, Sydney, Innsbruck, the Himalaya Zoo in Simla and Indianapolis.

As Director of the Animal Psychology Department at the University of Zurich, he gave fascinating lectures, enlivened by many practical examples, and he supervised numerous doctoral theses. Heini Hediger was a corresponding member of the Zoological Society of London and of the Senckenberg Natural History Society of Frankfurt. He also participated in the editorial boards of several scientific journals and books, *Zoo Biology* among others. Heini Hediger received many international honours for his services, including the Conservation Gold Medal from the Zoological Society of San Diego in 1974 'for dedication and service to the cause of wildlife conservation'. The award of the cultural prize of the city of Zurich with a laudatio of Konrad Lorenz in 1974 represented the crowning glory of his work.

In the world of zoos, he has changed animal husbandry profoundly through his publications, leadership and by founding the science of zoo biology. Upon retiring from the zoo world, his studies increasingly centred around the fundamental question raised by biology, namely the origin of life on Earth.

In 1996, in honour of Heini Hediger's leadership in zoo development, IUDZG initiated the 'Heini Hediger Award for Outstanding and Dedicated Service to IUDZG and the Zoo and Aquarium Profession'.

On Receiving the Heini Hediger Award

By George Rabb, Past Director of the Brookfield Zoo and Past President of the Chicago Zoological Society

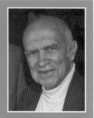

It was a very special honour to receive the award named in honour of Heini Hediger at the 1996 Annual Conference in Denver. At the Festkolloquium on his 80th birthday in Zurich in 1988, I had paid tribute to his pioneering studies in animal behaviour and to his influence thereby on thinking about human behaviour, social relations and architecture. Through my own and student studies of the wolf, Guinea baboon and okapi, I had early appreciation of the investigative insights on behaviour made by Heini Hediger and his students, such as Rudolf Schenkel and Hans Kummer.

Heini Hediger also set a great example for others who became engaged in the evolution of zoological parks as centres for research and for conservation by sharing his knowledge from the field and zoos through his numerous publications and by personal visits to their institutions. In retrospect, Heini Hediger brought science into the world of zoos, along with strong empathetic feelings for the animal subjects and objects of our attention. Central to his contributions were his observations on the behavioural dimensions of individual space.

He made the point that there is no absolute freedom for any creature. But he also maintained that we have the responsibility to foster the full expression of natural behaviours of animals whatever constraints are in place and whether the place is an apparently untrammelled landscape or seascape, national park or zoo exhibit. Relating to these thoughts, despite immediate conservation issues at the time in 1996, both Ulysses Seal and I felt it important to recognise Heini Hediger's single leadership and contributions to zoo biology by accepting the WAZA award in his name. I still feel so!

WAZA's Role Today and in the Future

What began as a simple network of zoos and a handful of zoo directors has evolved into an umbrella organisation for the worldwide zoo and aquarium community, which aims to help its constituency realise their full conservation potential. WAZA's mission is to be the voice of the worldwide community of zoos and aquariums and to act a catalyst for their joint conservation action. WAZA's plan for the future is best described by its Corporate Strategy, which sets out its course until 2020.

WAZA's Corporate Strategy Towards 2020

WAZA's vision: The full conservation potential of world zoos and aquariums is realised. This vision statement expresses WAZA's overall long-term goal of ensuring that the huge potential of zoos and aquariums throughout the world to contribute to species and habitat conservation and sustainability is fully realised.

WAZA's mission: WAZA is the voice of a worldwide community of zoos and aquariums and catalyst for their joint conservation action. This mission statement articulates WAZA's special role or 'niche' in achieving the vision as the global communication platform and representative for a major part of the world zoo and aquarium community; as well as the global catalyst for joint conservation action, business development, marketing, sustainability and membership.

The six strategic directions around which the Corporate Strategy is built are:

1. Develop and deliver WAZA's core conservation activities: A clear link between *in situ* and *ex situ* conservation is established and communicated. WAZA and its members are all engaged in conservation education and raising public awareness, capacity building, information exchange and technical training. All conservation work is based on sound scientific evidence, and the highest standards according to the WAZA *Code of Ethics and Animal Welfare* are applied. Institutions implement international environmental standards. The aquarium community is better integrated with the zoo community.

WAZA's Corporate Strategy Towards 2020

2. Develop and strengthen WAZA's external partnerships: Strategic partnerships with multilateral environmental agreements, intergovernmental organisations and international nongovernmental organisations are further developed; and strategic business partnerships are established with the corporate sector.

3. Increase WAZA's visibility and positive impact: Internal information flow is improved, external communications strengthened and a strong WAZA brand developed and promoted.

4. Improve and develop internal organisation, reflecting needs of institutional members and the associations: A well-defined and accepted WAZA Constitution – with clear roles, responsibilities and mandate – is developed. An appropriate organisational structure is evolved to pursue the Strategy and action plans based on it. Functional efficiency is maximised using the 'SMART' (Specific, Measurable, Achievable, Realistic, Time-bound) approach to all activities. Membership services are improved, meeting changing needs. Services and benefits for the aquarium community are defined and representation significantly increased.

5. Secure the financial growth and stability needed to implement the Strategy: WAZA's core budget is increased and its financial support base diversified. Specific financial provisions are made and capacity to catalyse or implement concrete activities developed.

6. Maintain and develop a culture of professional management and governance: Progress with implementing the Strategy is properly monitored – and feedback is acted on – to ensure that WAZA's strategic directions remain relevant, sound and on track. All constituents of the WAZA membership apply 'best-practice' in governance and management.

How will
a world-class zoo
and aquarium look like
in 25 years from now?

A medley
of predictions

Gunther Nogge,
WZO President 1994–1995 and Past Director of the Cologne Zoo

One of the resolutions of the first Environment Summit in Rio de Janeiro, Brazil, in 1992 was Agenda 21. *The 21st century is only 10 years old, but who speaks about its Agenda anymore? In the years following the Rio Summit, zoos developed their* World Zoo Conservation Strategy *and started to turn zoos into conservation centres, one was even renamed as a Wildlife Conservation Park. This process has now slowed down. Present economic constraints force zoos all over the world to set new priorities to attract enough people in order to survive economically. So, I look into the future a little concerned and I fear that 25 years from now zoos will be run by business and marketing people instead of scientists and conservationists, and they will have turned into entertainment parks with animals rather than conservation centres.*

Mark Stanley Price,
Senior Research Fellow at the University of Oxford

Such an institution will be financially secure as a valued part of the local community and society as relevant to the welfare of mankind in a challenging era.

Dave Morgan,
Executive Director of the African Association of Zoos and Aquaria (PAAZAB)

[I] hope that they will be facilities that seamlessly arch between free-ranging and ex situ *populations of animal species.*

Rick Barongi,
Director of the Houston Zoo

There will be far fewer zoological facilities but the ones that do survive will be totally immersive habitats for animals and visitors. The barriers will be virtually invisible allowing visitors to get very close to certain animals. New technology will capture all animal behaviour and provide a more stimulating and enriching environment for all the animal residents.

Every animal exhibit will have a direct connection to the wild habitat. The conservation messages will be a virtual reality display, 'transporting' the visitor to the natural habitats so they will be more inspired to care and help preserve what little biodiversity is left in the world.

Lee Ehmke,
Director of the Minnesota Zoo

Exquisitely detailed replicas of ecosystems, augmented by technological tools enabling 'virtual' connections in real time with the actual wild habitats being represented as well as the off-exhibit breeding/management areas being placed 'on-show'.

Mark Penning,
WAZA President 2010–2011 and Director of the uShaka Sea World Durban

Displays of complex ecosystems, showing how human populations can function within these as part of the system by maintaining a balance.

Joanne Lalumière,
Executive Director of the Granby Zoo

Zoos' role in educating more and more urbanised human populations in wildlife preservation will take on dimensions unsuspected today; accreditation will be considered by governments to issue permits; zoos will partner to 'own' or work with governments to manage sanctuaries.

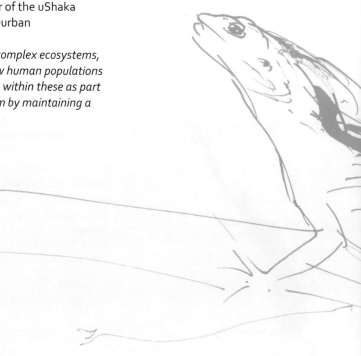

Chris West,
CEO of Zoos South Australia

Zoos metamorphose from physical sites doing some conservation to being headquarters of vigorous conservation organisations – accepted and funded.

Lena Lindén,
CEO of the Nordens Ark

All in situ *projects are supported and carried out by the local population and managed by the zoo. The only role of zoos and aquariums is conservation with the aim to achieve a sustainable population of an endangered species in the wild!*

Simon Stuart,
Chair of the IUCN Species Survival Commission

I think a world-class zoo or aquarium in 25 years time will have species conservation as its driving mission. In order to achieve this, the zoo will have extensive in situ *and* ex situ *conservation programmes focusing in particular on the species most at risk of extinction. Such institutions will address complex biological, social and economic issues in their* in situ *programmes, and will communicate the difficulties and complexities associated with achieving successful conservation to their public. Zoos will attempt to make their public more informed about conservation and much more supportive of it. They will resist efforts to dumb down communications.*

On ex situ *conservation, they will employ state-of-the-art techniques to rescue species from extinction and to rebuild numbers, as well as pioneering increasing numbers of successful reintroductions. Zoos have a unique constituency and they will educate and grow their local constituencies to become powerful political advocates to promote much-needed environmental policies that politicians are often unwilling to pursue for fear of unpopularity. Zoos will model sustainable lifestyles and institutions, and set an example to society of a sustainable future. Zoos will continue to enthuse both young and old alike, and should aspire to be in a leading position in species conservation worldwide. Zoos will work collaboratively with other sections of the conservation movement. Zoos should aim to grow their conservation budgets by one order of magnitude each decade. The global situation demands serious levels of ambition.*

Sally Walker,
Founder and Director of the
South Asian Zoo Association for
Regional Cooperation (SAZARC)

There will be traditional city zoos sized at a few acres to 250–300 acres with large naturalistic enclosures only far more advanced in their design so that both staff and visitors are sufficiently removed from the animals that there is no animal–human bonding that would interfere with releasing appropriate individuals in the wild, if required. The design would be such that visitors do not feel removed from the animals. With high-tech virtual devices, which manipulate sight, hearing, smell, touch, etc., visitors would participate in visits to forests, jungles, lakes, oceans, etc., so realistic and exciting that their intellect and emotions will be captivated. Subtle cues and dramas will convey the environmental crisis so that visitors will be touched in a deeper way and emerge from their visit, ready to work for conservation. Zoos will provide projects on site that visitors who want to make a difference can do immediately and take home the result to work on there.

There will be enormous non-traditional wildlife parks, sanctuaries, etc., where visitors are admitted only to do some work in a particular area (such as chop brush, turn over soil, study tiny animals in dead trees, record animal data as seen from strategically placed hides with CCTV in hundreds of places with a designated segment of forest, desert, water body, etc.). Powerful computers will create the work assignments that will be given to visitors on the basis of forms they will fill out, measuring their need to learn and their ability to do certain work. Visitors will develop an attachment to their experience that can be reinforced by follow-up 'visits' using virtual means and further tapping them for various work. Misbehaviour of any kind will not be tolerated but miscreants will not be thrown out of the facility. Instead, they will be sent to a special series of zoo exhibits, which will convey experiences that will change their attitudes and behaviour.

In 25 years, wildlife will have reduced to the extent that all people understand its value and look at wild animals the way people today look at the rarest, most beautiful and costly items. Zoos will attain the dignity, respectability and status of museums. People in all countries will pay heavily to see these animals and to learn about them. Zoo personnel from animal keeper to researcher to director will make frequent and regular public appearances on all forms of media and 'talk up' to their audience so that the public comes to understand the zoo process, which increases their respect.

All zoos will be members of WAZA. There will be no dysfunctional zoos. Zoos that do not come up to standard will be given training, a business plan and some time to improve. If they do not improve, the zoo will be closed and the animals will be sent anywhere in the world where they are needed for education, research, breeding and conservation, and will be appropriately treated.

Appendices

Appendix I

Members of the **Internationaler Verband der Direktoren Zoologischer Gärten** (International Association of Directors of Zoological Gardens) in 1936, participating in the second Annual Conference in Cologne (* denotes founding members of the Association in 1935):

- THEODOR ALVING, Copenhagen Zoo *
- OTTO ANTONIUS, Vienna Zoo *
- ALARIK BEHM, Skansen Zoo *
- RICHARD CLARKE, Bristol Zoo
- THOMAS GILLESPIE, Edinburgh Zoo
- HEINZ HECK, Munich Zoo *
- LUTZ HECK, Berlin Zoo *
- FRIEDRICH HAUCHECORNE, Cologne Zoo *
- MICHEL L'HOËST, Antwerp Zoo *
- JULIAN HUXLEY, Zoological Society of London *
- KOENRAAD KUIPER, Rotterdam Zoo *
- RODERIK MACDONALD, Philadelphia Zoo
- WILLIAM MANN, National Zoological Park Washington, DC
- RICHARD MÜLLER, Königsberg i. Pr. (Kaliningrad) Zoo *
- HERBERT NADLER, Budapest Zoo *
- KURT PRIEMEL, Frankfurt a. M. Zoo *
- MARTIN SCHLOTT, Breslau (Wroclaw) Zoo
- FRITZ SCHMIDT-HOENSDORF, Halle a. d. S. Zoo *
- KARL-MAX SCHNEIDER, Leipzig Zoo
- ARMAND SUNIER, Amsterdam Zoo *
- KAZIMIERZ SZCZERKOWSKI, Posen (Poznań) Zoo *
- KARL THÄTER, Nuremberg Zoo *
- ACHILLE URBAIN, Bois de Vincennes Zoological Park and Ménagerie du Jardin des Plantes Paris
- ADOLF WENDNAGEL, Basel Zoo *
- JAN ZABINSKI, Warsaw Zoo *

Appendix II

WAZA's *Code of Ethics and Animal Welfare*

Preamble

The continued existence of zoological parks and aquariums depends upon recognition that our profession is based on respect for the dignity of the animals in our care, the people we serve and other members of the international zoo profession. Acceptance of the WAZA World Zoo Conservation Strategy is implicit in involvement in the WAZA.

Whilst recognising that each region may have formulated its own code of ethics, and a code of animal welfare, the WAZA will strive to develop an ethical tradition which is strong and which will form the basis of a standard of conduct for our profession. Members will deal with each other to the highest standard of ethical conduct.

Basic principles for the guidance of all members of the World Association of Zoos and Aquariums:

viii. Assisting in achieving the conservation and survival of species must be the aim of all members of the profession. Any actions taken in relation to an individual animal, e.g. euthanasia or contraception, must be undertaken with this higher ideal of species survival in mind, but the welfare of the individual animal should not be compromised.

ix. Promote the interests of wildlife conservation, biodiversity and animal welfare to colleagues and to society at large.

x. Co-operate with the wider conservation community including wildlife agencies, conservation organisations and research institutions to assist in maintaining global biodiversity.

xi. Co-operate with governments and other appropriate bodies to improve standards of animal welfare and ensure the welfare of all animals in our care.

xii. Encourage research and dissemination of achievements and results in appropriate publications and forums.

xiii. Deal fairly with members in the dissemination of professional information and advice.

xiv. Promote public education programmes and cultural recreational activities of zoos and aquariums.

xv. Work progressively towards achieving all professional guidelines established by the WAZA.

At all times members will act in accordance with all local, national and international law and will strive for the highest standards of operation in all areas including the following:

1. Animal Welfare

Whilst recognising the variation in culture and customs within which the WAZA operates, it is incumbent upon all members to exercise the highest standards of animal welfare and to encourage these standards in others. Training staff to the highest level possible represents one method of ensuring this aim.

Members of WAZA will ensure that all animals in their care are treated with the utmost care and their welfare should be paramount all times. At all times, any legislated codes for animal welfare should be regarded as minimum standards. Appropriate animal husbandry practices must be in place and sound veterinary care available. When an animal has no reasonable quality of life, it should be euthanised quickly and without suffering.

2. Use of Zoo and Aquarium Based Animals

Where "wild" animals are used in presentations, these presentations must:

a. deliver a sound conservation message, or be of other educational value,
b. focus on natural behaviour,
c. not demean or trivialise the animal in any way.

If there is any indication that the welfare of the animal is being compromised, the presentation should be brought to a conclusion.

When not being used for presentations, the "off-limit" areas must allow the animal sufficient space to express natural behaviour and should contain adequate items for behavioural enrichment.

While the code focuses on zoos and aquarium based "wild" animals, the welfare of domestic animals, e.g., sheep, goats, horses, etc., in, e.g., petting zoos should not be compromised.

3. Exhibit Standards

All exhibits must be of such size and volume as to allow the animal to express its natural behaviours. Enclosures must contain sufficient material to allow behavioural enrichment and allow the animal to express natural behaviours. The animals should have areas to which they may retreat and separate facilities should be available to allow separation of animals where necessary, e.g., cubbing dens. At all times animals should be protected from conditions detrimental to their well-being and the appropriate husbandry standards adhered to.

4. Acquisition of Animals

All members will endeavour to ensure that the source of animals is confined to those born in human care and this will be best achieved by direct zoo to zoo conduct. The advice of the appropriate Species Co-ordinator should be sought before acquiring animals. This will not preclude the receipt of animals resulting from confiscation or rescues. It is recognised that, from time to time, there is a legitimate need for conservation breeding programmes, education programmes or basic biological studies, to obtain animals from the wild. Members must be confident that such acquisitions will not have a deleterious effect upon the wild population.

5. Transfer of Animals

Members will ensure institutions receiving animals have appropriate facilities to hold the animals and skilled staff who are capable of maintaining the same high standard of husbandry and welfare as required of WAZA members. All animals being transferred will be accompanied by appropriate records with details of health, diet, reproductive and genetic status and behavioural characteristics having been disclosed at the commencement of negotiations. These records will allow the receiving institution to make appropriate decisions regarding the future management of the animal. All animal transfers should conform to the international standards and laws applying to the particular species. Where appropriate, animals should be accompanied by qualified staff.

6. Contraception

Contraception may be used wherever there is a need for reasons of population management. The possible side effects of both surgical and chemical contraception, as well as the negative impact on behaviour, should be considered before the final decision to implement contraception is made.

7. Euthanasia

When all options have been investigated and the decision is taken that it is necessary to euthanise an animal, care will be taken to ensure it is carried out in a manner that ensures a quick death without suffering. Euthanasia may be controlled by local customs and laws but should always be used in preference to keeping an animal alive under conditions which do not allow it to experience an appropriate quality of life. Whenever possible a post-mortem examination should be performed and biological material preserved for research and gene conservation.

8. Mutilation

Mutilation of any animal for cosmetic purpose, or to change the physical appearance of the animal, is not acceptable. Pinioning of birds for educational or management purposes should only be undertaken when no other form of restraint is feasible and marking animals for identification should always be carried out under professional supervision, in a way that minimises suffering.

9. Research Using Zoo Based Animals

All zoos should be actively involved in appropriate research and other scientific activities regarding their animals and distribute the results to colleagues. Appropriate areas of research include exhibit design, observations, welfare, behaviour, management practices, nutrition, animal husbandry, veterinary procedures and technology, assisted breeding techniques, biological conservation and cryopreservation of eggs and sperm. Each zoo undertaking such research should have a properly constituted research committee and should have all procedures approved by a properly constituted ethics committee.

Invasive procedures designed to assist in medical research are not to be performed on zoo animals however the opportunistic collection of tissues during routine procedures and collection of material from cadavers will, in most cases, be appropriate.

The well-being of the individual animal and the preservation of the species and biological diversity should be paramount and uppermost in mind when deciding upon the appropriateness of research to be undertaken.

10. Release-to-the-Wild Programmes

All release-to-the wild programmes must be conducted in accordance with the IUCN/SSC/Reintroduction Specialist Group guidelines for reintroduction.

No release-to-the-wild programme shall be undertaken without the animals having undergone a thorough veterinary examination to assess their fitness for such release and that their welfare post-release is reasonably safeguarded. Following release, a thorough monitoring programme should be established and maintained.

11. Deaths of Animals Whilst in Care

Unless there are sound reasons not to do so, each animal which dies in captivity, or during a release to the wild program, should undergo post-mortem examination and have a cause of death ascertained.

12. External Wild Animal Welfare Issues

While this code of practice is designed for animals held within Zoos, Aquariums, Wildlife Parks, Sanctuaries, etc., WAZA abhors and condemns ill-treatment and cruelty to any animals and should have an opinion on welfare issues for wild animals external to its membership.

WAZA requires that:

- The taking of animals and other natural resources from the wild must be sustainable and in compliance with national and international law and conform with the appropriate IUCN policy.
- Any international trade in wild animals and animal products must be in compliance with CITES and the national legislation of the countries involved.

WAZA opposes:

- Illegal and unsustainable taking of animals and other natural resources from the wild, e.g. for bush meat, corals, fur or skin, traditional medicine, timber production.
- Illegal trade in wild animals and wild animal products.
- Cruel and non-selective methods of taking animals from the wild.

- Collecting for, or stocking of animal exhibits, in particular aquariums, with the expectation of high mortality.
- The use, or supply of animals for "canned hunting", i.e. shooting animals in confined spaces, or when semi tranquilised or restrained.
- Keeping and transporting of animals under inadequate conditions, e.g., the keeping of bears in confinement for extraction of bile, dancing bears, roadside zoos or circuses/entertainment.

WAZA and its members should make all efforts in their power to encourage substandard zoos and aquariums to improve and reach appropriate standards. If it is clear that the funding or the will to improve is not there, WAZA would support the closure of such zoos and aquariums.

This document was prepared on the basis of the 1999 Code of Ethics and the 2002 Code of Animal Welfare. It was adopted at the Closed Administrative Session of the 58th Annual Meeting, held on 19th November 2003 at San José, Costa Rica

Appendix III

WAZA's 2010 Bylaws

ARTICLE I

Name

The name of the organization shall be the **World Association of Zoos and Aquariums** (herein referred to as WAZA), which was re-founded after World War II in Rotterdam, The Netherlands on 24 September 1946 after the original founding in 1935 in Basel as the International Union of Directors of Zoological Gardens and later referred to as the World Zoo Organization.

Under the name of the **World Association of Zoos and Aquariums**, there exists an organization ("Verein") within the meaning of art. 60 *et seq.* of the Swiss Civil Code. The organization has its domicile in **Gland**, Switzerland and is based at the IUCN World Headquarters, 28 Rue Mauverney and has been founded for an undetermined period of time. The activities of the organization may be carried on throughout the world.

Only such assets as are owned by WAZA may be used to satisfy claims of creditors of the WAZA and assets of the members are specifically excluded.

ARTICLE II

Objectives

The objectives of WAZA shall be:

- to promote cooperation between zoological gardens and aquariums with regard to the conservation, management and breeding of animals in captivity, and to encourage the highest standards of animal welfare and husbandry;
- to promote and coordinate cooperation between national and regional associations and their constituents;
- to assist in representing zoological gardens and aquariums in other international organizations or assemblies;
- to promote environmental education, wildlife conservation and environmental research.

ARTICLE III

Membership

**Section 1.
General Requirements**

Members of WAZA shall be individual zoological gardens, aquariums, or organizations operating a number of such facilities, national and regional associations, and individuals willing to abide by the Bylaws and all other rules and regulations as defined in the various sections of this *Article III*. Failure to satisfy or adhere to the Bylaws, Code of Ethics and Animal Welfare and all other rules and regulations shall be sufficient cause for suspension or denial of membership. There is also a reasonable expectation that voting Member representatives will have the ability to attend Annual Conferences. Only members that have settled their yearly membership dues according to the provisions of Art. III, section 10 and 11 are eligible to vote.

Section 2.
Institution Members

Institution members shall be zoological gardens, aquariums or similar zoological institutions established and managed primarily for cultural, educational, scientific, and conservation purposes that are open to the public on a regular and predictable basis, as well as organizations operating a number of such facilities. An Institution member must also be either an accredited or full institutional member of its recognized regional or national association where such associations exist. Only the respective full time, paid chief executive, or other nominated senior executive of the institution or organization may serve as its official representative.

Section 3.
Association Members

Association members shall be organizations, both national and regional, whose primary members are zoological gardens, aquariums or similar zoological institutions. The purpose of these organizations is to support the vision, mission, and interests of their members and to establish standards and levels of cooperation between them. Association members may officially be represented by either an elected officer, or by a full time, paid executive.

Section 4.
Affiliate Members

Affiliate members shall be organizations that support the vision, mission and interests of WAZA. The official representative of an Affiliate member may attend the Annual Conference including Administrative sessions. Affiliate members may not be represented on Council or standing committees, but may be on other committees and working groups. They are not entitled to a vote and shall not be considered when determining a quorum.

Section 5.
Corporate Members

Corporate Members shall be individuals or entities that provide supplies or services to zoological gardens or aquariums and who support the vision, mission and interests of WAZA. The official representative of a Corporate Member may attend the Annual Conference including Administrative sessions. Corporate Members may not be represented on Council or standing committees, but may serve on other committees and working groups. They are not entitled to a vote and shall not be considered when determining a quorum.

Section 6.
Life Members

Life members shall be those who have retired as full time, paid executives of an Institution or Association Member, and who have served in that capacity at one or more Members for a period of no less than ten (10) years. Life members may not serve on Council or a standing committee, but may serve on other committees and working groups. They may also attend the Annual Conference including Administrative sessions. They are not entitled to a vote and shall not be considered when determining a quorum.

Section 7.
Honorary Members

Honorary members shall be those persons deemed worthy of such recognition as exemplified by their active support of WAZA's objectives. Only the Council may elect individuals to Honorary membership. Honorary members may not serve on Council or a standing committee, but may serve on other committees and working groups. They may also attend the Annual Conference, including Administrative sessions. They are not entitled to a vote and shall not be considered when determining a quorum.

Section 8.
Membership Process

An Institution, Association or Organisation desiring membership in WAZA must first be proposed as a candidate (in writing) to the Membership Committee by at least two voting members who are preferably from that country or region, but do not serve on the Membership Committee. Executive Director shall request candidates to submit appropriate information and associated materials for consideration by the Committee as required. The Committee will review the documentation for compliance with the WAZA's membership requirements. If approved by a simple majority of the Committee, the candidate will be recommended to the Council for approval.

If approved by a simple majority of the Council, information pertaining to the candidate shall then be published in the WAZA newsletter for a period not to exceed three (3) months. During that time, any voting member may express their objection or concern to Council regarding the candidate's qualifications. If no substantial objections or concerns are raised, the candidate shall be approved by Council and become a member of WAZA.

Candidates failing Committee or Council approval shall be so notified by the Executive Director. A denied candidate may not be proposed or reapply for membership for a period of one year. A denied candidate may, however, appeal in writing to WAZA's President for reconsideration.

In case the information and associated materials required for consideration by the Committee is not received within 24 months following the nomination, the membership process will be terminated and the candidate has to re-apply.

Section 9.
Membership Qualifications

All WAZA members have to sign WAZA's Code of Ethics and Animal Welfare, however qualifications vary between the membership categories of Institution, Association and Affiliate and Corporate.

Criteria to be considered for Institution membership shall include:

- proper animal husbandry and veterinary care;
- participation in coordinated species management programmes;
- participation in conservation activities;
- participation in relevant scientific studies;
- compliance with applicable national and international legislation;
- the maintenance of record systems and cooperation with studbook and species support programmes;
- environmental education programmes;
- support of national and international conservation programmes;
- endorsement of WAZA's Code of Ethics and Animal Welfare;
- membership in a recognized regional and national association as appropriate;
- the ability to be officially represented at Annual Meetings.

Associations are required to provide documentation that includes information regarding their mission, Bylaws, general operation, overall function and the ability to be officially represented at the Annual Meeting.

Affiliates are required to provide information similar to Associations, however, representation at the Annual Conference is at the Affiliate's discretion.

Section 10.
Membership Dues and Services

The Council shall establish annual membership fees and services for Institution, Association, Affiliate, Life, Honorary and Corporate members. Membership fees are assessed on an annual basis and are due on the 31st January.

Section 11.
Suspension and Expulsion of Members

Non-compliance with the WAZA Code of Ethics and Animal Welfare or failure to pay annual membership fees within ninety (90) days of the date of the invoice may be cause for suspension and potential expulsion from WAZA.

Upon the written recommendation of the Membership Committee, WAZA's Finance Committee Chair, or a written concern submitted to the President by the Executive Director or by any Member regarding an ethics violation, the Council may for cause, suspend or terminate membership. An Institution Member will also lose its membership in WAZA if it fails to maintain either an accredited status or a regular membership in its recognized regional or national association. Such action requires a two-thirds (2/3) majority vote of the Council. Cause shall be failure to satisfy *Article III, Sections 1* and *10* of these Bylaws.

Any member affected by this action shall be so notified by certified mail. If desired, the member may appeal in writing to the WAZA's President for reconsideration by the Council. Only the Council may reinstate. This appeal must occur within sixty (60) days following notification and may include a request to appear before the Council. The member shall be classified as "suspended" during any process of appeal. The Council must act upon the appeal prior to the next annual Administrative Session. Failing or waving the right of appeal shall cause the Council to notify the membership of the expulsion.

Section 12.
Termination of Membership by the Member

Any member may withdraw from WAZA by giving written notice. The information must be sent by October 1st of the previous year in question to the Executive office and the notice will be sent by the Executive office to the Membership Committee.

In order to avoid the obligation to pay the annual membership fees for the upcoming fiscal year, a member's written notice of its resignation must be received by the WAZA Executive Office 90 days (i.e. latest October 1st) prior of that fiscal year.

Membership can only be terminated on an annual basis.

The WAZA Executive Office will acknowledge termination of membership in writing within 30 days of notice of termination. The annual membership fee is non-refundable and any outstanding dues must be paid accordingly despite termination.

ARTICLE IV

The Council

Section 1.
Composition and Responsibilities

The Council shall consist of the President, the President elect, who shall function as the Vice-President, and seven other members, one of which shall serve as chairman of the Finance Committee, and another as Chairman of the Membership Committee. The officers and council members shall represent geographic regions according to a composition as determined by Council. This composition shall be determined as follows:

- Region I shall be North America, consisting of Canada and the USA.
- Region II shall be Europe and the Middle East.
- Region III shall consist of Mexico, Central America and the Caribbean, South America, Africa, Asia excluding the Middle East, and Oceania (Australia, New Zealand, Papua New Guinea and the Pacific Islands).

The Council shall be entrusted with the general direction and operation of the Association and may adopt such rules as necessary, within the limits of the Bylaws, to transact business. Each member of Council shall have a single vote. In the event of a tie in voting, the President shall cast the deciding vote. The Council shall be reflective of the Association's demographics and no member of the Council shall receive compensation for his or her services. The Association's full time, paid Executive shall serve as an ex-officio member of the Council without the right to vote.

Section 2.
Officers

The Council officers shall be the President and President elect.

Section 3.
Terms of Office

Terms for each elected member of the Council shall consist of two years respectively. These terms shall be understood to begin and end at the conclusion of the applicable Annual Conference, or until their successors are elected, whichever comes later. The President elect will progress to President by ratification of the membership. Other members of Council may be eligible for re-election, but may not serve more than three full terms consecutively.

In the event of a vacancy occurring in the position of President, the President elect will assume the vacated position. If the vacancy is that of the President elect, then the Council and the Nominating Committee shall determine a replacement from the remaining Council Members. The appointees will complete the current term and then serve their own term.

In the event of a vacancy with a Council member, following discussion with the region concerned, the Nominating Committee will submit the name of a suitably qualified candidate to Council for ratification. This individual will serve as a member of Council until the next respective election.

Section 4.
Nomination to Council

A candidate for election to the Council must satisfy the official representative requirement of *Article III, Sections 2* and *3* of these Bylaws for a period of at least five (5) years and be a voting member in good standing for at least three (3) years at time of election. Individuals already serving as officers or members may be considered for re-election if applicable. Candidates for election to Council shall be proposed through whatever method deemed appropriate by the colleagues of the respective regional or national association. Individuals considered for nomination to Council must provide a letter of support, which indicates that the candidate will be available for a given period of time and that all travel expenses will be provided.

To be considered, the name of the proposed candidate must be submitted in writing to the Chairman of the Nominating Committee by the last day of January of the designated election year.

The number of candidates to be considered for election to Council shall consist of at least one more than the number of vacancies. If not enough candidates are nominated by WAZA Members as of the established deadline, the Nominating Committee shall have the authority to identify additional candidates from the respective region.

The Nominating Committee shall submit its recommendations in writing to the Council prior to its mid year meeting during a designated election year. Following approval by the Council, a mail ballot shall be distributed to the voting members in accordance with *Article VII, Section 2* of these Bylaws. Optionally, electronic voting via e-mail ballots or other web-based procedures may be offered.

Section 5.
Selection of Officers

Candidates for President-elect are selected by the Nominating Committee in consultation with Council. The President-elect is expected to progress to the office of President.

ARTICLE V

Meetings

Section 1.
The Annual Conference

WAZA shall hold an Annual Conference during each calendar year, which shall include an Administrative Session (the Annual meeting of the Association, to conduct formal and legally-required business). The purpose of the Conference shall be to share zoological information, promote cooperation among members, accept the reports of committees and working groups, approve WAZA's financial statements and budget, and to conduct other business as required. The Administrative Session shall be subject to normal parliamentary procedures and such other rules and regulations as established by the Bylaws. The Annual Conference will not normally be held in the same country more than once in every three years.

Delegates to the Annual Conference must satisfy all appropriate conference fees as a condition of participation in the daily events and other conference activities. The same financial responsibility shall be expected of non-voting representatives attending the annual meeting and conference activities. Additional participants who are associated with Members may attend the conference only after being invited by the President. Their participation shall also be subject to a predetermined conference fee.

Non members or other organizations may be invited by the Council to attend WAZA's conferences as observers. Observers may attend all sessions except Administrative, participate in discussions, and will be subject to a predetermined conference fee. They are not entitled to a vote and shall not be considered when determining a quorum.

Section 2.
Council and Other Meetings

The Council shall hold at least two meetings a year, and as many others as deemed necessary for the purpose of transacting Council business. Council meetings are normally closed meetings. The Council may occasionally request that additional WAZA related meetings convene for more specific purposes. These meetings shall also be subject to normal parliamentary procedures if applicable.

Section 3.
Quorum

A quorum for the annual Administrative Session shall consist of a simple majority of the voting members registered as delegates at the Annual Conference. A majority of the Council or committee members shall constitute a quorum for the transaction of Council or committee business. Quorums are not required for the actions of working groups.

Section 4.
Notice of Meetings

All voting members shall be notified, at least sixty (60) days in advance of all WAZA related meetings. This notice shall specify the date(s), location, purpose of the meeting and where applicable, who should attend. If the notice is for the Annual Conference, it shall also request the submission of proposed agenda items for consideration by the Council for the Administrative Session.

ARTICLE VI

Standing, Other Committees and Working Groups

Section 1.
General Responsibilities

The Council, at its discretion, may form standing committees as well as other committees and working groups and shall appoint their members for the purpose of addressing specific issues and to satisfy the objectives of WAZA. Only standing committees shall be formally identified and defined in these Bylaws.

These groups shall function in accordance with the Bylaws and be subject to normal parliamentary procedures if applicable. All members of standing committees shall be voting members of WAZA. The President shall serve as an ex-officio member of all committees and working groups except the Nominating Committee, without the right to vote.

Section 2.
Membership Committee

The Membership Committee is responsible for reviewing applications for membership or changes in membership status in accordance with *Article III* of these Bylaws. The Committee shall also assist in promoting membership growth in WAZA.

The Committee shall consist of a Chairman and six other members who shall be reflective of the WAZA's demographics. The President shall, when necessary, appoint the Chairman and other members to serve for a term of two (2) years respectively with the possibility of appointment for one additional term. All those serving on the Membership Committee shall be ratified during the annual Administrative Session.

Section 3.
Nominating Committee

The Nominating Committee is responsible for verifying that individuals proposed for consideration as elective members of Council are appropriately qualified in accordance with Article IV of these Bylaws. It is also the Committee's responsibility to identify the appropriate number of qualified candidates required to satisfy the elective process for each region. In addition, the Nominating Committee shall determine which Council members are interested in being considered for the position of President Elect, and inform Council prior to the Council's vote.

The Committee shall consist of five members, including the Immediate Past-President, one current member of the Council (excluding the President), and three others who are appointed by the incoming President for a period of two years and are reflective of WAZA's demographics. If possible, the Immediate Past-President shall serve as Chairman of the Committee. The three appointees may not succeed themselves or be considered for nomination. All those serving on the Nominating Committee shall be ratified during the annual Administrative Session.

Section 4.
Ethics and Welfare Committee

The Ethics and Welfare Committee is responsible for monitoring the Organization's members and their compliance with the Code of Ethics and Animal Welfare in accordance with *Article II* and *Article III*, *Section 9* of these Bylaws. It also has the responsibility of investigating alleged member violations of this Code and the subsequent reporting of its confidential findings to Council.

The Committee shall consist of a Chairman and six other members who shall be reflective of WAZA's demographics. The President shall, when necessary, appoint the Chairman and other members to serve for a term of two (2) years respectively with the possibility of appointment for one additional term. All those serving on the Ethics and Welfare Committee shall be ratified during the annual Administrative Session.

Section 5.
Finance Committee

The Finance Committee is responsible for advising the Council in matters pertaining to the WAZA's financial health and welfare. This includes monitoring its operational revenues and expenses, accounts and investments, and addressing any issues of a financial nature as may be assigned by Council. The Committee also has the responsibility to initiate an annual audit by selecting an appropriate entity qualified to conduct such activities and processes.

The Committee shall consist of a Chairman and three other members who are knowledgeable and experienced in matters of budgeting and financial accountability. The Chairman and one member of the Committee shall be members of Council. The other two members shall be non Council members. The President shall appoint the Chairman and other members to serve for a term of two (2) years respectively with the possibility of appointment for one additional term. All those serving on the Finance Committee shall be ratified during the annual Administrative Session.

Section 6.
Appointment to Committees

Only the President may appoint members to serve on standing committees. The President shall also have the authority to select chairpersons for standing and other committees. Members of non-standing committees shall serve without term limits.

ARTICLE VII

Voting and Elections

Section 1.
Voting Members

Voting members shall be those who officially represent Institutions or Associations pursuant to the rules of membership as defined in *Article III*, *Sections 2 ,3, 10 and 11*. In the event of a tie, the President shall be called upon to cast an additional vote. If a ballot is returned partially unmarked or contains in that portion a non-valid number of votes or has to be regarded invalid due to unclear marking, the unmarked or otherwise invalid portion shall be considered a non-vote, and will be deducted from the total number of eligible votes for the respective portion that serve as the basis for a simple majority.

Section 2.
Ratification of Officers and Election of Council Members

Ratification of officers and election of Council shall occur by mail ballot. Printed ballots containing the names of the proposed candidates and appropriate information relevant to each shall be distributed to the voting members of WAZA. Ballots must be returned to the WAZA's Executive office by the specified date. Optionally, electronic voting via e-mail ballots or other web-based procedures may be offered. Candidates shall be elected according to the number of votes they receive from the voting membership. In the event of a tie, the respective candidates shall be subject to a second balloting process. The names of the President and President-elect shall appear on the ballot for the purpose of ratification by the membership.

Section 3.
Ratification of Standing Committee Members

Individuals to be considered for service on the Membership, Nominating, Ethics and Welfare and Finance Committees, shall have their names placed before the membership, by the President, for ratification during an Administrative Session. Approval by a simple majority of the voting members present, which shall include the use of proxies, is required for ratification. Voting shall consist of any means deemed appropriate by the President.

Section 4.
Approval of Resolutions

Proposed resolutions must be submitted in writing to the Council within forty-eight (48) hours of an Administrative Session. Formal approval of a proposed resolution shall require both a simple majority vote of the Council and a simple majority of the voting members present, which shall include the use of proxies. Voting shall consist of any means deemed appropriate by the President.

In an emergency, a resolution may be adopted by a two-thirds (2/3) majority of the Council. The voting members must ratify the emergency resolution during the next Administrative Session for approval.

Section 5.
Motions

During an Administrative Session, the President may request that a voting member of the delegation move an agenda or other item, for approval. This shall be accomplished through a motion and a second to accept or approve, followed by a simple majority vote for approval, which shall include the use of proxies. Voting shall consist of any means deemed appropriate by the President.

Section 6.
Amendment of the Bylaws

WAZA's Bylaws may from time to time be amended. Proposed amendments or revisions to WAZA's Bylaws may be initiated by the Council or recommended by a petition of no less than ten percent (10%) of the voting members. Amendments submitted by petition do not require the approval of the Council, but must be submitted at least ninety (90) days prior to the next Annual Conference. Proposed amendments shall be placed on the agenda for discussion during the Administrative Session. If approved by a simple majority of the quorum, the Council shall distribute printed ballots by mail to all voting members of WAZA, which must be returned to WAZA's Executive office by the specified date. Approval of the proposed amendment requires a two-thirds (2/3) majority of the votes received.

Section 7.
Special Elections

The Council at its discretion may initiate special elections for the purpose of electing members to Council or amending the Bylaws. These elections shall be conducted in accordance with *Article VII, Sections 1, 3 and 7* of these Bylaws.

Section 8.
Proxies

Proxies may be utilized for the purpose of voting during the Administrative Session. Their use is restricted to resolutions, ratification and various motions in accordance with *Article VII, Sections 4, 5 and 6*. A proxy shall be assigned to a voting member in the form of a recognized proxy statement. Voting members holding proxies shall identify themselves prior to voting activities. Proxies shall not be considered when determining a quorum.

ARTICLE VIII

Parliamentary Authority

Section 1.
Parliamentary Procedure

The Council shall be the final authority in interpretations of the Bylaws and other rules and regulations of WAZA.

ARTICLE IX

Governing Law, Language

Section 1.
Governing Law

The Council relationship among the members and the acts of the various bodies of WAZA shall be governed by these statutes, which shall in all respect be regarded as made under and governed by the laws of Switzerland.

Section 2.
Language

The English text of these statutes shall prevail over all others.

Adopted and put into force by the General Assembly at its 65th Administrative Session, Köln, Germany, 21st October, 2010.

Appendix IV

Locations of WAZA's Annual Conferences from 1935 to 2012

1935	Basel, Switzerland	1978	Leipzig, Germany
1936	Cologne, Germany	1979	Warsaw, Poland
1937	Munich, Germany	1980	Pretoria, South Africa
1938	Amsterdam, the Netherlands	1981	Washington, DC, USA
1946	Rotterdam, the Netherlands	1982	Rotterdam, the Netherlands
1947	Basel, Switzerland	1983	Melbourne, Australia
1948	Paris, France	1984	Flevohof, the Netherlands
1949	Copenhagen, Denmark	1985	Calgary, Canada
1950	London, UK	1986	Wroclaw, Poland
1951	Amsterdam, the Netherlands	1987	Bristol, UK
1952	Rome, Italy	1988	Stuttgart, Germany
1953	Antwerp, Belgium	1989	San Antonio, TX, USA
1954	Copenhagen, Denmark	1990	Copenhagen, Denmark
1955	Basel, Switzerland	1991	Singapore
1956	Chicago, IL, USA	1992	Vancouver, Canada
1957	Rotterdam, the Netherlands	1993	Antwerp, Belgium
1958	Frankfurt, Germany	1994	São Paulo, Brazil
1959	Copenhagen, Denmark	1995	Dublin, Ireland
1960	Cologne, Germany	1996	Denver, CO, USA
1961	Rome, Italy	1997	Berlin, Germany
1962	San Diego, CA, USA	1998	Nagoya, Japan
1963	Chester, UK	1999	Pretoria, South Africa
1964	Taronga, Australia	2000	Palm Desert, CA, USA
1965	Berlin, Germany	2001	Perth, Australia
1966	Colombo, Sri Lanka	2002	Vienna, Austria
1967	Barcelona, Spain	2003	San José, Costa Rica
1968	Pretoria, South Africa	2004	Taipei, Taiwan
1969	New York, NY, USA	2005	New York, NY, USA
1970	East Berlin, Germany	2006	Leipzig, Germany
1971	Prague, Czech Republic	2007	Budapest, Hungary
1972	Amsterdam, the Netherlands	2008	Adelaide, Australia
1973	Tokyo, Japan	2009	St. Louis, MO, USA
1974	Basel, Switzerland	2010	Cologne, Germany
1975	Colorado Springs, CO, USA	2011	Prague, Czech Republic
1976	Caracas, Venezuela	2012	Melbourne, Australia
1977	Vienna, Austria		

Appendix V

What Does Our Union Stand For?

By Armand Sunier, IUDZG Past President. A paper presented at the 1952 Annual Conference in Rome.

For the administrators of a Zoological Society difficulties sometimes arise from the fact that there are many intermediate stages between a *bona fide* zoological garden, an animal show on a fairground or the show of a dealer who is constantly trading in animals.

This is apt to lower the standing of the *bona fide* zoo in the eyes of at least part of the public and of the authorities. Now it is certainly superfluous to emphasise in this meeting that it is necessary for the *bona fide* zoos to have the goodwill of the public and the support of the authorities. When I came to the zoo in Amsterdam 25 years ago, nearly everyone in the Netherlands who was interested in the protection of animals, the prevention of cruelty of animals, or the protection of nature, was opposed to zoological gardens. Now, by stressing for years the great value of a good zoo as an indispensable educational institution, the care that is taken of the live animals in such a zoo, both by zoologists and veterinarians, the survival and breeding in zoos of species extinct in nature, as for example, the European Bison, we have a least succeeded in justifying the existence of a good zoo in the eyes of those who were formerly against it. I tried to make it understood that the reason for the existence of a zoo is its social importance as an educational institution.

In this connection we employ in our zoo a staff of young biologists who are able to conduct parties through the gardens, telling them in an entertaining way all kinds of interesting things about our animals. We thereby succeed annually in offering to tens of thousands of our visitors recreation in a form that has a certain mental, intellectual and aesthetic value. As our colleague Crandall put it once in a letter to me: "We try to give our public as much information as possible in a chocolate coating."

Amongst the things that most easily shock the feelings of many persons, are questions connected with the capture of and the trade in wild animals. I have always felt, therefore, that a *bona fide* zoo, in order not to lessen its standing in the eyes of the public and the authorities, must shun as far as possible, connections with animal dealers. Of course, every zoo must from time to time buy and sell animals or at least exchange them for others. Without this a representative collection of living wild animals could hardly be maintained. But, in my opinion, it is entirely wrong for the trade in animals to become one of the means by which the zoo tries to make money. It is true that the character of an institution like a zoo depends more upon the way in which its money is spent than on the means by which it is earned. If the money is spent only to attain the ideal aim of a real zoo, that is to give the public as much interest in and as much knowledge of live animals as possible, all should be well. However, it will still prejudice a zoo greatly in the eyes of a considerable part of the public and of the authorities if the money to attain the ideal aim is partly earned by traffic in wild animals. Most people nowadays have a feeling that a zoo should be of a standing very different from that of an animal dealer.

The question of zoos and of the traffic in wild animals was discussed at a technical meeting of the International Union for the Protection of Nature in the Hague in September 1951. At this meeting, I was asked to formulate a proposal concerning this matter, that could be brought before the session which has been held recently in Caracas. The text of this proposal, which has been accepted by Mr. Jean-Paul Harroy, reads as follows:

"It is desirable that in all countries the importation of animals belonging to species which are protected in their natural habitat should be prohibited

a. unless it has been definitely established that the exportation of such animals from their country of origin has been carried out under completely legal conditions.

b. unless the animals are destined for a scientific or municipal institution, non-profit-making and expressly recognised as such by the government of the importing country, and not for an animal dealer."

It would be necessary in all countries to prohibit the exhibition to the public or the possession of a species which are protected in their natural habitat unless the owner is a scientific or municipal institution, non-profit-making and expressly recognised as such by the government of the country where it is established.

I have already drawn your attention to the fact that species of wild animals extinct in nature can survive. To emphasise the importance of this question I should like to read a note I gave to Mr. Jean-Paul Harroy, at his request, to be brought before the General Meeting of the International Union for the Protection of Nature held at Caracas a few days ago.

"Everywhere in the world, nature is inevitably becoming more deeply modified by the constant incursions of the human population. In the very near future it will certainly no longer be possible to preserve all the natural surroundings to a sufficient extent to assure the survival in complete liberty of all the species threatened by extinction."

It is well to remember that the secret of political success has always been the art of not striving after the impossible. The essential is the survival of the threatened species. But in the future, and above all in the case of the larger mammals, it will often be absolutely impossible to conserve a species in its natural surroundings. For such a species it will be necessary to create a new and more appropriate habitat, where it will be able to survive in semi-captivity and where the very high mortality obtained in the natural habitat can be reduced by man taking care to assure its food, the necessary veterinary attention and all other protection of which it stands in need. It is undeniable that the sudden granting of political independence in a tropical territory, of which one has seen several recent examples, can constitute a very real danger to the survival of certain particularly vulnerable species. That is why it is a matter of urgency that several species should be considered as being in the nature of a prized possession of immense value.

So, for example, an appropriate habitat, where all the necessary care can be given to the animals, must certainly be found for the orang-utan and put at the disposal of an international institution provided with adequate resources. Without this, this species will certainly soon become extinct first in Sumatra, then in Borneo.

At the Paris meeting of our Union in 1948, Mr. Jean Delacour pointed out that our zoos can help in the protection of rare animals and of animals that are in danger of extinction

a. by not trafficking in these animals.
b. by breeding them.

Indeed in future all zoos will have more and more to reckon seriously with the facts that:

i. a well run zoo is an increasingly necessary, indeed indispensable educational institution of great social importance.
ii. the problem of saving many species of wild animals from extinction becomes rapidly more and more urgent.

Therefore, in the near future, the only excuse for the existence of a zoo will be that it fulfils adequately its educational task, which is wholly comparable with that of a museum, and that it contributes to the preservation of species of wild animals.

Adequately to fulfil its educational task and also, of course, to be able to give its live animals the food and care they need, a zoo must have a staff of experts who are capable, by reason of their scientific studies, of dealing with all aspects of animal welfare.

From this it will be evident that an institution like a zoo, primarily educational in character, acquires a scientific importance as, by the work of its scientists, data of scientific value concerning the behaviour, feeding, reproduction, pathology, &c. of the animals are collected daily. Besides this, the zoo offers, of course, an opportunity to many persons not belonging to its staff to study live animals in all respects from an intellectual or from an aesthetic point of view. Moreover the live animals of the zoo continually yield valuable material for many institutions of scientific research.

I have already mentioned that we, in our gardens, employ a number of young scientists to give the public as much information as possible in an entertaining and interesting way. In this connection I may tell you that with the able help and by the application of these young biologists I was able to realise this summer in our zoo a day-dream of mine, and one that I made mention of for the first time in a publication entitled "The Zoological Garden of To-day and of To-morrow" which appeared on the occasion of the centenary of the Royal Zoological Society Natura Artis Magistra in 1938.

This day-dream is a small building in which the public space is kept in twilight and the walls of which a number of large circular ground glass panes display projections of small and microscopic live animals 40 to 150 times enlarged. These projections of live animals interest the public far more than the cinematographic projection of microscopical photographs. I have brought you a map with photographs of this new department of our gardens to which we have given the popular name of Artis 'micro-world'. To this map are attached two specimens of a new kind of label we have introduced in our gardens and which may interest you.

The principles I have put before you are the same that underlie the wording of Article 2 of the Constitution of our Union which I was allowed to draw up for you and which was accepted by you in 1949. They also underlie the text of Article 1 of the Law of the Royal Zoological Society Natura Artis Magistra of Amsterdam. They enable us to answer the question: What does our Union stand for?

It stands for the maintenance of zoos that are essential non-profit-making, educational institutions, like museums; that do their utmost to give the public, in a pleasant way, the benefit of as much knowledge of and interest in live animals as possible and that, moreover, promote the protection and preservation of the world's species of wild animals.

Finally, I think I may in this connection refer once more to the need of a zoo, and also of our Union, having the sympathy of the public and the support of the authorities.

At the moment our Union certainly has both these advantages, but should it deviate from the principles I have mentioned, it could easily lose them. This could happen if persons are elected to membership of the Union whose *bona fide*s do not fulfil exactly the requirements laid down in Article 3 of our Constitution.

In the eyes of the public and of the authorities, the standing, and therewith the authority of our Union, will always depend exclusively upon the quality and certainly never upon the quantity of its members.

Appendix VI

The Changing Role of Zoos in the 21st Century

By William Conway, Past President and General Director of the Wildlife Conservation Society.
Keynote speech at the 1999 Annual Conference in Pretoria.

What is happening outside zoos?

If you knew that wildlife is declining so fast that the end of animal acquisitions for zoos is in sight, as a zoo director what would you do? How would this affect your zoo's capital improvement plans? It's education programs? It's conservation efforts?

In a general way, we all do know that human populations are increasing and that wildlife is disappearing. But, we also hear that many things are getting better. Outside Africa and India, human life expectancies have dramatically increased and large numbers of people are much better fed and cared for today than fifty years ago. Even population growth is slowing. To many of us, most environmental problems appear solvable. But what are the facts insofar as wildlife, hence zoos, are concerned?

Twenty-five percent of all birds have been driven to extinction in the last 200 years. Eleven percent of the remaining birds, 18% of the mammals, 5% of the fish and 8% of terrestrial plants are now seriously threatened with extinction (Barbault and Sastrapradja, 1995). Almost all big animals are in trouble; storks and cranes, pythons and crocodiles, great apes (in fact, most of the primates), elephants and rhinos. Ninety percent of the black rhinos have been killed in the past eighteen years and one third of the world's 266 turtle species are now threatened with extinction.

More to the point is the plight of tropical forests. Most terrestrial species are found in such forests but only about 7.5 million km2 of tropical evergreen forests remain. If their deforestation continues at the same rate as it did between 1979 and 1989, the last tropical forest tree will fall in 2045, but the rate is increasing (Terborgh, 1999). Moreover, tropical forest wildlife is under much more pressure than even these figures suggest. The bushmeat harvest in the Brazilian Amazon is now estimated at 67,000 to 164,000 metric tons per year and this kill is dwarfed by that in the tropical forests of Africa, now more than one million metric tons each year (Robinson et al., 1999). Field scientists are reporting a new phenomenon – the "empty forest" (Redford, 1992).

At the same time, we are besieged by reports of acid rain, decreases in tropical forest rainfall, ozone depletion, global warming, coral bleaching, phytoplankton blooms, cancer epizootics in fishes and such extraordinary declines as the dramatic worldwide loss of amphibians (Myers, 1999).

Our growing herds and flocks of domestic animals have become a plague to wildlife, devastating habitat and spreading disease – anthrax, rinderpest, distemper. Bovine tuberculosis has spread from domestic cattle to wild buffalo, thence to lions, cheetahs, kudus and baboons here in South Africa while it threatens wood bison in Canada.

Over forty percent of Earth's total terrestrial photosynthetic productivity is now being appropriated by people (Vitousek et al., 1986), who also consume 25–35% of the primary productivity of the ecosystems of the continental shelves (Roberts, 1997) and use 54% of all run-off in rivers, lakes and other accessible sources of fresh water (Postel et al., 1996). Along with the assault of our domestic animals upon the land and its wildlife, we are casually distributing an unrelenting stream of wild exotics, ranging from European boars in California to red deer in Argentina – to say nothing of what has happened in the famously disastrous situations in New Zealand, Hawaii and Australia. The destructive effect of alien species of plants and animals wins a place second only to man's in the U.S. (Eisenrink, 1999; Kaiser, 1999; Stone, 1999). In California's San Francisco Bay, for example, an average of one new species has been established every 36 weeks since 1850, every 24 weeks since 1970, and every 12 weeks in the last decade.

Although it is true that the rate of human population growth has slowed, nearly one billion people are being added to the population every 12–13 years, far more in real numbers than 50 years ago; the compounding effect of a smaller rate on a larger base. This is occurring mostly in countries where Earth's biodiversity is greatest.

The life expectancy of wildlife is not better than 50 years ago. It is incomparably worse. Nevertheless, not one national government on Earth has made preservation of its environment a top priority. Only major environmental catastrophes seem likely to win humanity's attention and loss of species will rarely qualify as catastrophic.

Only 4% to 6% of the terrestrial realm and 0.5% of the marine realm are under any sort of wildlife protection (Freese, 1998). For most people, wildlife conservation is a luxury and no great scientific breakthrough can be expected to stop extinction simply, no immunization, quarantine or wonder drug.

Species-specific requirements of big predators are especially sobering. For example, a female jaguar in Peru requires a minimum of 20 sq. kms. for herself and her cubs while a pair of harpy eagles needs about 50 sq. kms. The two million hectare Manu Reserve system supports only 10 families (60 individuals) of giant otter (Terborgh, 1999). A single Indian tiger requires a standing herd of about 700 axis deer to produce sufficient food on an annual basis (U. Karanth, pers. comm.).

What does this mean for zoos?

Jack Welch, CEO of General Electric, has observed: "When the rate of change on the outside exceeds the rate of change on the inside, the end is in sight." In the outside world of wildlife and nature, which we represent to our hundreds of millions of visitors, the rate of change far exceeds the internal zoo response. Like so much of nature, zoos also face extinction – unless they are able to change. Despite the good work of our WZO planning task force, zoos have been overtaken by the speed of wildlife extinction. To survive and fulfil their obligations to society, they must become proactive conservation organizations, not living museums, and they must do it now.

What is happening on the inside? Zoo disconnects.

Inside the zoo world there are marvellous new exhibits, fascinating breeding successes, improved education programs, ever better curatorial management and veterinary treatment, and much more, but also an alarming series of disconnects; of zoo priorities, unrelated to the priorities of wildlife, unresponsive to change – ultimately unrelated to the future of zoos. How many of us have focused our education efforts on those people in the strongest positions to affect the future of the wildlife we exhibit? For the most part, we target our conservation education on children and other non-decision makers in a process too slow or too far away to address the extinction crisis in which we now live. Our efforts to inform law-makers and government authorities are usually low-key or non-existent. Our campaigns are more likely to be for a new gorilla exhibit than for the existence of gorillas at all.

In fact, some zoo people feel that conservation messages might give their visitors a "negative experience"; that only positive messages should be provided – an idea reminiscent of the auto industry's old argument against seatbelts! Almost all of us participate in the "professional conservation conspiracy", hiding from our patrons and donors how bad things really are for the fear that they will withdraw their support in the face of a seemingly hopeless cause. But conservation action is not negative, nothing could be more positive or more exciting. The WCS donors and trustees that actually get out to see what is happening to nature are now our strongest supporters. Wildlife conservation is destined to be among the main adventures as well as challenges of the 21st century. Our heroes will save wonderful wild creatures and beautiful wild places against the odds. And if these heroes are not recognized as such, it will be our fault.

Zoos seldom participate in species or habitat restoration. In the U.S., organizations such as the National Audubon Society, the International Crane Foundation and the Peregrine Fund are re-establishing auk, tern and puffin colonies, cranes, eagles, falcons and even small island birds – not zoos. Where we have acted, as in the restoration of the California condor, black-footed ferrets, Wyoming toads and the long-ago American bison reintroductions; our potentials are clear. But the fact is that our current propagation programs for vanishing species are usually too small and subdivided to provide either the number of animals or the scientific samples needed for sound maintenance, restoration or research.

Most of us contribute little financial support to *in situ* wildlife conservation programs, either to research, education or training. Few of us operate parks or reserves or participate in their management and founding. In 1992, less than 325 *in situ* conservation projects were being supported by AZA zoos, 85% by the Wildlife Conservation Society. Today, however, the number exceeds 650, only about 50% by WCS. If this trend continues, I believe that zoos could become the primary non-governmental field conservation organizations. Yet, zoos seldom act collectively to address national or international conservation issues. Few zoo biologists are experienced with the fundamental ecological problems that must be addressed in reserve management and few zoos are the main places our public turns to for wildlife information, beyond the baby-bird-out-of-the-nest variety. We need to train a new breed of zoo people.

How did such disconnects come about. Quite naturally, of course. Most zoos were created as educational and cultural institutions for their local communities and were meant to help convey the gifts of biological literacy and enjoyable recreation. Saving wildlife was not much in the minds of their founders. But even today most of our conservation education efforts are leisurely, uninvolving and indirect, not significantly different from museums exhibiting fossils. We seldom teach the unsettling facts of population biology and even more rarely related it to our visitors' own behavior. Such "conservation" as zoos attempt is mostly a passive and generalized advocacy, not directly affecting issues or locations. But that was then. This is now. Then, the conservation crisis was not anticipated, perhaps understandably. Now it can not be ignored, not understandably.

How can we modify our vision?

The prospects for humanity, and for wildlife, depend upon stopping habitat destruction and human population growth at a sustainable level wherein people can still attain a rewarding and desirable quality of life and where the beauty and diversity of life itself is cared for and nurtured. The immensity of the challenge is highlighted by E.O. Wilson's estimate that it would take two more planet Earths to support the current global population in a lifestyle now common in North America.

Nevertheless, we can not address the future from a foundation of pessimism. What kind of world might informed biologists allow themselves to hope for? What kind of zoo vision? Current guesstimates suggest that human numbers will level-off during the next 50 years at about 8.8 billion.

Unduly optimistic or not, I propose that WZO embrace a vision of the future wherein human effects upon the environment have been tethered and considerable wildlife remains; certainly not as rich or abundant as today's wildlife but with substantial diversity and biomass and numbers of more or less wild ecosystems – and that zoos work to make this vision become reality.

We will have to sustain smaller than normally viable wildlife populations in reserves and parks – and living with us, as part of human-dominated landscapes, lives and livelihoods. Zoos must help save fragments both *ex situ* and *in situ* (Conway, 1999).

Thus the 21st century zoo must be redesigned as a hedge against biotic impoverishment; a time machine buying continuance for faltering wildlife populations; a corridor of care between parks and reserves; and, more than ever, humanity's primary introduction to wildlife, promoter of environmental literacy and recruiting center for conservationists.

The room available to wildlife in nature's remainders will be too small to long sustain viable populations of creatures which require big spaces without help, to say nothing of migratory forms whose specialized resources for the seasons of their lives are far apart. The dominant creatures will have to be managed on a species by species basis, with the main focus upon the "landscape species" whose broader ecological needs help define the ecological systems we will seek to sustain (Lambeck, 1997; Robinson, 1999; Conway, 1999). Their requirements will provide the matrix for the biological and cultural advocacy essential for the survival of the thousands of smaller niches necessary to the lesser creatures that share parts of their landscape.

Such a concept requires expert and unswerving population management of big predators and larger landscape species whose size and needs could carry the seeds of destruction for their newly restricted ecosystems. It will require monitoring at zoo levels of intensity; the curatorial and veterinary management that today is lavished only on zoo animals in SSPs and EEPs.

The arts and sciences of translocation, reintroduction and habitat restoration will have to be further developed in a species-specific way, and by organizations accustomed to dealing with multi-species problems and free of political special interests; by zoos – if they can grow up to it.

In this connection, much of what has been said about reintroduction of naïve captive-bred animals in the "wild" is less and less relevant. Future opportunities will be as much "introduction" as "reintroduction", for available habitats may be greatly altered, lacking their original complements of predators and plants as well as the same dangers or the same resources.

Fundamentally, the saving of wildlife is a social process burdened with widely disjunct cultural values. Ultimately, it depends less upon "how?" than upon "why?" The existence and charisma of living animals themselves are the best answers. It is a special role for zoo education.

What changes must zoos and aquariums make?

All this is fine for generalities, I suppose, but what about specifics? How can zoos and aquariums respond to such a vision? What changes must they consider? Three seem self-evident: First, because wildlife habitats are disappearing, most zoo animals will have to be collectively managed in closed populations, so zoo programs must be planned, above all, to sustain long-term viable populations. SSPs and EEPs must be expanded. (The average SSP is now only 143 specimens.) And this also means that, except for zoo relationships with parks and reserves, other zoo priorities must be secondary. The next break-through exhibits may be given over to the propagation of endangered species, perhaps with zoo-goers on a pay-per-view basis, their support going to the protection of parks and reserves. The "Noah's Ark paradigm" will, inevitably, come back into its own as we seek to save animal "seeds" for habitat restoration and partner with reserves and other kinds of conservation programs.

Second, maintenance of high species diversity in zoo collections creates diseconomies of scale because of the need for many costly species-specific protocols. Thus collection planning must focus more on specialization with animals having compatible requirements, and far more upon international collaboration.

Third, to sustain interbreeding populations of the species they exhibit and contribute to the survival of under-sized park populations, zoos will have to make a larger commitment to the sciences of applied ecology, assisted reproduction and population management. Nevertheless, to propose captive propagation as a primary solution to the loss of major habitats and wildlife biomass is to trivialize conservation science and the knowledge we have of it – a topical treatment for a tumor.

Some zoos are responding to these concerns. The Denver Zoo, and several others, has built conservation contribution machines where visitors may give cash towards the preservation of chosen species. Modified parking meters have long been used by various collections to generate support for rain forest preservation. Perhaps charging a conservation percentage at zoo entrances is an appropriate option. New York's Bronx Zoo has opened a major complex entirely focused upon Congo Basin conservation wherein the exhibit admission fees may be voted toward the cost of specific Congo conservation projects. It is expected to generate nearly US$ 1 million each year.

The acquisition of "zoo reserves" by coalitions of zoos has been proposed (Conway, 1998) as a way of protecting more habitat, providing local incentives for conservation and also providing for monitored off-takes for zoo exhibits. It may work best with small, short-generation species, as a constructive alternative to costly breeding programs.

Beyond the direct implications of the extinction crisis for zoo collections is the zoo's potential roles in directly saving nature. Two loom large:

- Reaching and advising major decision-makers, and all the others we can, about the nature of biological limitations and specific conservation issues.

- Directly helping to sustain wildlife in nature – or what must pass for nature – in the years ahead. Helping to sustain wildlands, reserves and species, especially those which have lost their habitats, is our greatest potential service to society. More than 115 parks and reserves protecting about 61 million ha. have resulted from WCS's *in situ* conservation efforts.

Surely it is time for every new wild animal exhibit to answer three central questions positively: If this exhibit were not built, would wildlife be hurt, helped or unaffected? Will it provide for the continuity of its inhabitants? Will it contribute to species preservation in nature?

A zoo exhibit's contribution to conservation may be financial, scientific, educational or through propagation – but it may no longer ignore the extinction crisis. It is past time for zoos to stop arguing that exciting children in New York or Tokyo about the plight of gorillas in Cameroon or Congo is responsive conservation. It is too indirect, too slow, too far away and too unlikely to affect the real issues.

Today's exhibit techniques allow us to present wildlife situations as never before, visualize places our visitors could never go, bring nature into our zoos in real time, interpret the beauty and ecology of the creatures we see at any magnification, and offer animal population simulations under any scenario… if we can sustain the animals. Otherwise all this becomes paleontological.

Where else should zoos be focusing? We pay too little attention to Third-World zoos located on the front lines of the Earth's most biodiverse habitats. Seventeen of the 20 largest cities on Earth will soon be located in the Third-World. Their zoos are struggling to survive. We must help them to make a difference where it counts, where the wildlife is.

And it is essential to take advantage of the extraordinary opportunities in communication presented by the Internet. By this I do not mean inter-zoo communication alone, as important as that surely is, but communication with conservation decision-makers; with people in or near wildlife areas and current dilemmas who need information, whether they know it or not. The admirable networking of CBSG in this regard deserves stronger zoo support. Indeed, the work of the Specialist Group already provides the international zoo community with much of whatever positive voice it has. Unless WZO itself is able to coordinate its actions and programs globally, it will not only be ineffective in international conservation but also in winning the support needed for the growth and continuance of zoos.

Thus the Zoo's vision for the 21st century should be to become proactive wildlife conservation care-givers and intellectual resources; to step out beyond our fences by aiding parks and reserves; to sustain animals which have lost their habitats and conduct campaigns to restore them – and to provide from our collections as many key species as possible to be the stimulus and centerpieces of conservation efforts around the world.

Such commitments must come to be recognized as of a different order than those we make in humanity's other preservation efforts. We need make little sacrifice to save art, literature and music – no long term commitment to a future that might really affect our individual pieces of the economic pie. But to save nature we must sacrifice the opportunities to consume or destroy it. And instant gratification for its saviors is rare.

Far more than most of the art and literature of our time – to say nothing of our more trivial entertainments – wildlife science, education and preservation resonate with moral purpose and importance, with wholesome aims and prospective significance. Inevitably, this makes the contemplation of wildlife conservation uncomfortable.

To become proactive conservation organizations, engines of conservation, is a powerful and inspiring role for zoos in the next century. "But", as one of my colleagues put it, "zoos and aquariums were not designed to be conservation organizations." The question is, can they be?

References:

- Barbault, R. and S. Sastrapradja. 1995. Generation, maintenance and loss of biodiversity In: Heywood, V.H. and R.T. Watson (eds.), Global Biodiversity Assessment. Cambridge University Press, Cambridge, pp 193–274.
- Conway, W. 1998. Zoo reserves; a proposal. AZA Annual Conference Proceedings, Tulsa, Oklahoma, 54–58.
- Conway, W. 1999. Linking zoo and field, and keeping promises to dodos. 6th International Endangered Species Breeding Conference, Cincinnati, in press.
- Eisenrink, M. 1999. Biological invaders sweep in. *Science*, 285:1834–1836.
- Freese, C.H. 1998 Wild Species as Commodities: Managing Markets and Ecosystems for Sustainability. Island Press, Washington, DC. 319 pp.
- Kaiser, J. 1999. Stemming the tide of invading species. *Science*, 285:1836–1841.
- Lambeck, R.J. 1997. Focal species: A multi-species umbrella for nature conservation. *Conservation Biology*, 11:849–856.
- Myers, N. 2001. The biodiversity outlook: Endangered species and endangered ideas. In: Shogren J.F. and J. Tschirart (eds.), Social Order and Endangered Species Preservation. Cambridge University Press, Cambridge, pp. xxv–xxxvi.
- Postel, S.L., G.C. Daily and P.R. Ehrlich. 1996. Human appropriation of renewable fresh water. *Science*, 271:785–787.
- Redford, K. 1992. The empty forest. *BioScience*, 42:412–422.
- Roberts, C.M. 1997. Ecological advice for the global fisheries crisis. *T.R.E.E.*, 12:35–38.
- Robinson, J. 1999. Biodiversity conservation at the landscape scale. Unpublished document. Wildlife Conservation Society, Bronx. 88 pp.
- Robinson, J.G., K.H. Redford and E.L. Bennett. 1999. Wildlife harvests in logged tropical forests. *Science*, 284:595–596.
- Stone, R. 1999. Keeping paradise safe for the natives. *Science*, 285:1837.
- Terborgh, J. 1999. Requiem for Nature. Island Press, Washington, DC. 234 pp.
- Vitousek, P.M., P.R. Ehrlich, A.H. Ehrlich and P.A. Matson. 1986. Human appropriation of the products of photosynthesis. *BioScience*, 36:368–373.

Appendix VII

Conservation of Nature: A Duty for Zoological Gardens

By Kai Curry-Lindahl, Past Director of the Nordic Museum and Skansen.
A paper presented at the 1964 Annual Conference in Sydney.

A modern zoological park has great obligations. As a working museum institution exhibiting living animals and, if possible, their habitats, it should be able to give the public cultural recreation combining both instruction and pleasure. It is of extreme importance and a great responsibility that this task is fulfilled in the right way, so that the public's interest is directed toward a wide perspective. An animal in its enclosure is doing more than simply presenting itself and its characteristics. It is a means of providing a broader understanding of the role that particular species play in a biocommunity. Such a goal can only be reached successfully by a zoo through a thorough understanding of living nature, the inter-relationship of animals and their environment. Therefore, research or at least full awareness of scientific progress in various ecological and zoological fields is an inevitable part of the work of the modern zoo. But there are also other scientific investigations that all zoos are morally obliged to undertake, because they are in a unique position to do so. Much ethological and pathological research as well as some biological observations can only be carried out in zoos, since living and dead material of certain species are available for such studies only in zoos. Mr. van den Bergh (1963) has already stressed the importance of this kind of work in volume 4 of the Yearbook.

Yet another important duty for each zoo is to work actively for the conservation of nature. One may say that a zoo is parasitizing on the animals it exhibits, because they are used to attract the public. The least we can do in return is to pay back a part of that debt to the animals. This can be done through careful, intelligent propaganda as to the value and importance of preserving wild animals in their natural habitats on different continents of this still interesting world of ours. We shall consider the animals exhibited in zoos as ambassadors for their wild relatives on savannahs, in forests and along the sea shore. But their success as ambassadors depends on us. It is true that an animal in its enclosure to a certain extent speaks for itself, but exhibitions, labels, guides, brochures, talks, press releases, should always reflect the theme of conservation.

If the zoos of the world, which together have an annual attendance of about 140 million visitors, all worked along these lines a tremendous amount could be done in educating the public to realize that conservation of nature is much more than something that aims at the preservation of animals because they are beautiful or amusing. People must be guided to understand that conservation of nature is necessary for the survival of man himself.

The four main functions of a zoological garden or an aquarium should be conservation, research, education, and public recreation. Few other museum institutions other than zoos have greater opportunities to achieve such a programme – extension of knowledge of nature conservation, particularly of ecological principles, amongst such a vast public. Zoos are especially well-equipped to do this as they are, in the main, open-air museums which can arrange permanent and temporary exhibits in cooperation with living nature itself.

I have put conservation and research as first items in a zoo's programme, deliberately because they form the base on which exhibits and other educational activities must rely.

Unfortunately, conservation principles are still far from being understood owing to lack of information during various stages of education, in schools as well as universities. This is a basic barrier that prevents the development of an appreciation of the need for conservation among people of all kinds of professions, as well as among political rulers.

Even if every country has its own separate problems in the field of conservation one can say that the methods to make people understand what conservation of nature and natural resources is about are very similar.

First of all, it must be stressed that it is necessary to develop an understanding of conservation because in the long run it is necessary for human existence. This is not speculation. If conventional food items for human beings are not replaced in the future by synthesized food, it will be impossible to produce sufficient proteins for the growing human populations, unless drastic measures of conservation and restoration of nature are undertaken.

This is only one aspect. There are many others. Preservation of natural habitats and their flora and fauna play an important part in fulfilling the need for human recreation, which is of social importance. Science requires natural communities of wild animals and plants for basic research. The landscape, the diversity of habitats, the various species of plants and animals, all belong to the cultural heritage that the generations of to-day are obliged to pass on intact to those who follow after us.

The idea of conservation is relatively new. It was born as a sensible reaction to man's destruction of nature and living natural resources. It has evolved through decades and its basic concepts have changed from what was at first purely sentimental concern about nature to the essentially practical approach of today. This evolution of attitude has unfortunately been slow in relation to the tremendously rapid destruction of the world's natural resources. Now, we have reached the desperate stage that necessitates an entirely new deal for the utilization of the Earth, for the sake of our own existence, if nothing else. This seriousness of the situation has become accentuated – in fact it is an emergency.

How are we to make human populations of different social and educative levels, as well as different age groups understand the problems and realize that conservation of nature is really necessary?

Education in schools is an important means, but it must in the long run be preceded by an alteration in the individual's approach to nature. Traditionally, people have regarded nature with its plants and animals as an enemy. Parent's instruction of their children emphasizes this attitude. Here, a zoological garden can do a lot to help and alter the climate of opinion by giving a large number of people basic information in a pleasant form.

The value of national parks and natural reserves as a means for conservation education should not be underestimated. It is most important that people should be encouraged to visit such areas and let them see for themselves, not only the wildlife of their own country, but also the enjoyment it brings to visitors from foreign countries. Here again, zoos can stimulate people to go out of doors to discover living nature for themselves.

If zoological gardens are to be efficient means of understanding nature conservation, it is essential that they should not be considered by different groups of people as antagonistic to conservation and animal protection. Unfortunately, such opinions are often expressed. Sometimes they are justified, but mostly they are based on misunderstanding and a subjective attitude that ignores the true nature of serious zoological gardens and what they are able to do positively for wild animals.

It is also important that zoos try to explain in different ways what ecology is. We must let the general public understand the complex inter-relations of living organisms and their environment, emphasizing inter-specific relations. Therefore, ecology ought to be presented as a scientific discipline in connection with all development planning and exploitation of renewable natural resources, especially in densely populated countries, where unwise use of land and water may have catastrophic results for the whole biocommunity including man.

There are already too many examples of man's industrial or agricultural projects that have failed completely economically and have only reduced a landscape to a ruin, where there was formerly a flourishing, highly productive area in its natural state. The reason why such disastrous schemes have been launched and even partly initiated or fully developed is often because in the first instance biologists were not consulted or that their advice, based on sound ecological data, was neglected. Therefore, biological and ecological viewpoints must be taken into consideration and be treated with the same amount of respect as purely economic arguments. In fact, ecology in this connection is a synonym with conservation of nature, and conservation can well be interpreted as applied ecology or human bio-economy.

It is a notable task for institutions such as zoological gardens, visited annually by many millions of people, to contribute to a better understanding of how our own environment functions.

Some zoo-people may object to the ideas I have tried to express. They may find such a philosophy and even more, such a programme beyond the scope of a zoo. But animals belong to nature's renewable resources and are, together with flora, water and soil, the basis of our existence. Surely zoos should be able to stress constantly, in a psychologically stimulating way, that animals are just part of the picture, that their habitats are as important as themselves. Thus, comparisons will easily be made with man's own situation. The visitors must be guided to understand that the management of the Earth's natural resources is essential for the well-being of human populations, and that it must be based on the modern concept of nature conservation which, from a human point of view, is a wise, long-term utilization of nature's renewable resources. The aim must be to achieve a biological balance between man's demands on the one hand and the extent to which nature can keep us with them on the other. Human needs are not confined to economic values alone; the yield produced by nature is also valuable for recreational, cultural and scientific considerations.

Several zoos have already done magnificent work for conservation, research and education. In this context, such institutions as the New York Zoological Society with its closely allied Conservation Foundation, the Zoological Society of London, the Société Royale de Zoologie d'Anvers and the Frankfurt Zoo must obviously be mentioned. Recently, the link between the IUCN (International Union for the Conservation of Nature and Natural Resources) and IUDZG have been strengthened through IUCN's Survival Service Commission and the Zoo Liaison Committee. This cooperation was further extended by the zoos and conservation symposium held at the Zoological Society of London in June 1964.

An increasing number of zoos are beginning to be aware of the need to support the conservation of nature. But work in this direction must be done whole-heartedly and not simply to try and justify the fact that some zoos consume many more animals than they produce.

To work for conservation is an obligation every serious institution must undertake if it exhibits living animals for museums or recreational purposes. Only by having such an attitude and undertaking such work can the existence of a zoological garden be fully justified.

Index

A...
Aalborg Zoo | 147
Adelaide Zoo | 38, 141
Admissions Committee | 10, 28
African Association of Zoos and Aquaria (PAAZAB) | 41, 85, 157
Amphibian Ark (AArk) | 15, 70, 93–96, 108
Amsterdam Zoo | 26, 38, 57, 63, 75, 148, 162
Antwerp Zoo | 18, 28, 30, 38, 42–43, 64, 87, 112, 137, 162
Apenheul Primate Park | 82
Association of Zoos and Aquariums (AZA) | 41–42, 108, 127, 135, 140, 187, 191
Associations Committee | 42, 44
Auckland Zoo | 85

B...
Badham, Molly | 46
Barongi, Rick | 7, 141, 158
Basel Zoo | 38, 104, 150, 152, 162
Benchley, Belle | 46
Berlin Zoo | 18–19, 38, 46, 68, 112, 162
Berne Animal Park | 46, 150, 152
Bonner, Jeffrey | 96
Brand, David | 38
Bristol Zoo | 7, 30, 92, 162
British and Irish Association of Zoos and Aquariums (BIAZA) | 41, 42
Bronx Zoo | 7, 134–135, 189
Brookfield Zoo | 117, 154
Building a Future for Wildlife: The World Zoo and Aquarium Conservation Strategy (WZACS) | 92–93, 99, 148
Byers, Onnie | 7, 106

C...
Calgary Zoo | 38
Chester Zoo | 7, 38, 93, 102, 148
Chicago Zoological Society | 70, 107, 154
Code of Ethics and Animal Welfare | 14, 31, 138, 141, 155, 163, 169–171
Cologne Zoo | 7, 38, 84, 87–88, 140, 144, 157, 162
Columbus Zoo | 135
Committee for Inter-Regional Conservation Cooperation (CIRCC) | 13, 84–86, 115
Committee for Population Management (CPM) | 44, 85–86, 108
Conference of Directors of Zoological Gardens of Central Europe | 19–21
Conservation Committee | 14, 90, 92
Convention on Biological Diversity (CBD) | 15–16, 60, 97, 99–100
Convention on Migratory Species (CMS) | 15, 60, 97–98
Convention on the International Trade in Endangered Species of Wild Fauna and Flora (CITES) | 16, 72–73, 100, 166
Conway, William | 2–3, 13, 47, 65, 82, 107, 119, 126, 150, 184, 187–189, 191
Copenhagen Zoo | 7, 19, 38, 45, 162
Crandall, Lee | 27, 180
Crudi, Lamberto | 21
Curry-Lindahl, Kai | 124–125, 192

D...
Daman, Frederic | 30–31, 38, 43–44, 137
De Boer, Leobert | 7, 43, 82, 85, 87, 150
Dennler, William | 52
Der Zoologische Garten | 20
Detroit Zoological Society | 93
Dick, Gerald | 1–2, 6, 15, 84, 91, 93, 97, 125, 131, 133, 147
Dollinger, Peter | 7, 14, 55, 72–73, 118, 150
Durrell, Gerald | 47–48

E...
Edinburgh Zoo | 31, 38, 43, 87, 137, 162
Education Committee | 57, 146–147, 149
Ehmke, Lee | 7, 158
Ethics and Welfare Committee | 141, 175–176
European Association of Zoos and Aquaria (EAZA) | 41–42, 82

F...
Fisher, Lester | 38, 47, 76, 79
Flesness, Nate | 7, 83, 116–118
Foose, Tom | 83, 107–108
Frankfurt Zoo | 7, 20, 23–24, 38, 66–67, 93, 195
Frankfurt Zoological Society | 20

G...
GaiaPark | 82
Gipps, Jo | 92
Global Conservation Coordinators' Committee | 85, 108
Global Species Management Plan (GSMP) | 86, 108
Granby Zoo | 158
Gray, Jenny | 7, 128
Grzimek, Bernhard | 23, 66–67, 153

H...
Hagenbeck Animal Park | 132
Harrison, Bernard | 52, 55, 85
Heck, Heinz | 21, 162
Hediger, Heini | 13, 38, 50, 83, 126, 150–154
Heinroth, Katharina | 46
Hering-Hagenbeck, Stephan | 7, 132
Hoessle, Charles | 47, 136–137
Houston Zoo | 7, 158

I...

International Association of Directors of Zoological Gardens | 10, 21, 22, 33, 62, 162
International Species Information System (ISIS) | 13, 43–44, 51–52, 77, 83, 87, 107, 116–118, 137
International studbook | 11–12, 15, 70, 85–86, 107–108, 112–115
International Union for Conservation of Nature (IUCN) | 2, 7–16, 60–61, 63, 65–66, 68, 70, 72–74, 77, 79, 81–82, 85–88, 92–94, 97, 99, 101–115, 117, 143, 148, 159, 166, 168, 195
International Year of Biodiversity | 16, 100
International Zoo Educators' Association (IZE) | 13–16, 43, 57, 100, 145–149
International Zoo Yearbook | 113–114
IUCN Species Survival Commission (SSC) | 6, 8, 11–13, 68, 70, 72, 74, 77, 79, 81, 85–88, 93–94, 97, 99, 103, 105–109, 111, 113–115, 117, 159, 166
IUCN/SSC Amphibian Specialist Group | 70, 94, 108
IUCN/SSC Conservation Breeding Specialist Group (CBSG) | 7–8, 12–13, 43–44, 59, 70, 74, 79, 81–83, 85–88, 93–94, 96–97, 105–111, 113–114, 117, 190

J...

Japanese Association of Zoos and Aquariums (JAZA) | 41–42, 131–132
Jersey Wildlife Preservation Trust | 110
Julin, Henning | 147
Junhold, Jörg | 4, 7, 38, 59

K...

Karsten, Peter | 33, 38, 50, 87
Khao Kheow Open Zoo | 88
Klös, Heinz-Georg | 18, 38
Krantz, Palmer | 38
Kuiper, Koenraad | 24, 26–27, 162

L...

Labuschagne, Willie | 7, 25, 38, 50, 85, 118, 150
Lacy, Robert | 7, 106–107
Lalumière, Joanne | 7, 158
Lang, Ernst | 38, 104
Leipzig Zoo | 7, 38, 59, 95, 122, 162
Lensink, Bart | 75
Lincoln Park Zoo | 7, 38, 47–48, 76
Lindén, Lena | 7, 94, 159

M...

Mallinson, Jeremy | 48, 110, 150
Marketing Committee | 59
Marton-Lefèvre, Julia | 7–8, 61, 97, 111
McAlister, Ed | 38, 52, 141
McCann, Colleen | 7, 134
McGregor Reid, Gordon | 6, 38, 48, 70, 96, 150
McKeown, Stephen | 7, 148
Metro Washington Park Zoo | 57
Meyer-Holzapfel, Monika | 46
Minnesota Zoo | 87, 108, 117, 158
Morgan, Dave | 7, 85–86, 157
Mottershead, George | 38, 102, 148
Munich Zoo | 7, 21, 121, 162

N...

Naples Zoo | 46
National Foundation for Research in Zoological Gardens | 43, 82
National Zoological Gardens Pretoria | 38, 50, 85
National Zoological Park Washington, DC | 162
Neugebauer, Wilbert | 75
New York Zoological Society | 27, 195
Niekisch, Manfred | 7, 23
Nogge, Gunther | 7, 38, 50, 84, 87, 140, 150, 157
Nordens Ark | 94, 159

O...

Olney, Peter | 7, 12, 113–114

P...

Pechlaner, Helmut | 54
Penning, Mark | 7, 38, 40, 91, 99, 158
Peters, Chris | 149
Philadelphia Zoo | 38, 162
Prague Zoo | 38
Priemel, Kurt | 10, 17–21, 34, 38–39, 162

R...

Rabb, George | 7, 13, 47, 70, 82–83, 93, 150, 154
Ramsar Convention on Wetlands | 15, 97, 99
Rawlins, Colin | 38, 104, 115
Reed, Theodore | 31, 107
Resenbrink, Han | 148
Reuther, Ronald | 29–30
Reventlow, Axel | 38, 45
Riverbanks Zoo and Garden | 38
Rotterdam Zoo | 7, 24, 26–27, 30, 38, 76, 125, 164
Rübel, Alex | 7, 38, 52, 55, 90, 151

S...

Saint Louis Zoo | 126, 136
San Diego Wild Animal Park | 54
San Diego Zoo | 38, 46, 102
San Francisco Zoo | 29
Sausman, Karen | 7, 14, 38, 46–47, 54, 150
Schroeder, Charles | 38, 102
Seal, Ulysses | 13, 47, 70, 82–83, 87, 106, 108, 110, 114, 116–117, 150, 154
Seifert, Siegfried | 38
Shelly, Freeman | 38
Sheng, Sherry | 57
Simón Bolivar Zoo | 88
Skansen | 7, 124, 162, 192
South Asian Zoo Association for Regional Cooperation (SAZARC) | 41–42, 160
Stanley Price, Mark | 7, 157
Strahan, Ronald | 29–30
Stuart, Simon | 7, 70, 159
Stuttgart Zoo | 75
Sunier, Armand | 10, 26–27, 38, 63, 123, 162, 180

T...

Taronga Zoo | 29
The Living Desert | 38, 46–47
Turning the Tide: A Global Aquarium Strategy for Conservation and Sustainability | 98
Twycross Zoo | 46

U...

United Nations Decade on Biodiversity | 9, 16, 98, 100
uShaka Sea World Durban | 38, 91, 99, 158

V...

Van Dam, Dick | 30, 38, 76, 125
Van den bergh, Walter | 7, 28, 38, 64, 112, 192
Verband Deutscher Zoodirektoren (VDZ) | 18, 20, 41–42
Veselovský, Zdeněk | 38
Vienna Zoo | 7, 19, 73, 105, 120, 140, 143, 149, 162

W...

Walker, Sally | 7, 42–44, 49, 150, 160
Washington Park Zoo | 57
Wenner, Claire | 46
West, Chris | 7, 94, 159
Wheater, Roger | 7, 31–32, 38, 43–45, 48, 50, 87, 137, 150
Wilcken, Jonathan | 7, 85–86
Wildlife Conservation Society | 7, 88, 93, 117, 119, 126, 134–135, 184, 187, 191
Windecker, Wilhelm | 38
World Zoo Conservation Strategy (WZCS) | 8, 13, 81–84, 86–87, 92, 108, 110, 123, 146, 148, 157, 163
Wylie, Stephen | 52

Y...

Yamamoto, Shigeyuki | 7, 130
Year of the Frog | 15, 94–95
Year of the Gorilla | 15, 98

Z...

Zippel, Kevin | 93, 96
Zoo and Aquarium Association Australasia (ZAA) | 41–42
Zoo Future 2005 | 13, 57, 59, 86–87
Zoo Liaison Committee | 103, 105–106, 113, 195
Zoological Society of London | 7, 11, 38, 82, 93, 104, 113, 115, 148, 153, 162, 195
Zoological Society of San Diego | 153
Zoos South Australia | 94, 159
Zoos Victoria | 128
Zurich Zoo | 38, 90, 126, 150–152

Participants at the 2011 Annual Conference in Prague (© Tomáš Adamec)

77 Years:
The History and Evolution of the World Association of Zoos and Aquariums 1935–2012

Laura Penn, Markus Gusset and Gerald Dick

World Association of Zoos and Aquariums
WAZA | *United for Conservation*®

Title:
77 Years:
The History and Evolution of the World Association
of Zoos and Aquariums 1935–2012

Authors:
Laura Penn, Markus Gusset and Gerald Dick

Publisher:
World Association of Zoos and Aquariums (WAZA)
Executive Office, Gland, Switzerland

Layout and design:
Michal Stránský, Staré Město, Czech Republic

Print:
Agentura Bravissimo, Znojmo, Czech Republic

Copyright:
© 2012 World Association of Zoos and Aquariums (WAZA)
Executive Office, IUCN Conservation Centre,
Rue Mauverney 28, 1196 Gland, Switzerland

Cover illustration:
kindly donated by Wolfgang Weber (www.wildlife-artist.de)

Drawings:
Zdeněk Tománek

ISBN: 978-2-8399-0926-6